Particles and Universes of Cosmos Theory

Geometric Unification of Coupling Constants, Masses and Structure
Separation of Fermion Masses into Sequences
Sequence-Based Particle Structure
Particle Universe Shells
Particle Universe Interiors
Vector and Scalar Boson Spectrums
Separation of Higgs and Other Scalar Bosons
Geometry of the Gravitation Constant G
Quantization of Universe Mass-Energy
Proof of Gambol Theory

Stephen Blaha Ph. D.
Blaha Research

Pingree-Hill Publishing
MMXXIV

To Margaret

Some Other Books by Stephen Blaha

SuperCivilizations: Civilizations as Superorganisms (McMann-Fisher Publishing, Auburn, NH, 2010)

All the Universe! Faster Than Light Tachyon Quark Starships & Particle Accelerators with the LHC as a Prototype Starship Drive Scientific Edition (Pingree-Hill Publishing, Auburn, NH, 2011).

Unification of God Theory and Unified SuperStandard Model THIRD EDITION (Pingree Hill Publishing, Auburn, NH, 2018).

The Exact QED Calculation of the Fine Structure Constant Implies ALL 4D Universes have the Same Physics/Life Prospects (Pingree Hill Publishing, Auburn, NH, 2019).

Passing Through Nature to Eternity ProtoCosmos, HyperCosmos, Unified SuperStandard Theory (Pingree Hill Publishing, Auburn, NH, 2022).

HyperCosmos Fractionation and Fundamental Reference Frame Based Unification: Particle Inner Space Basis of Parton and Dual Resonance Models (Pingree Hill Publishing, Auburn, NH, 2022).

The Cosmic Panorama: ProtoCosmos, HyperCosmos, Unified SuperStandard Theory (UST) Derivation (Pingree Hill Publishing, Auburn, NH, 2022).

God and and Cosmos Theory (Pingree Hill Publishing, Auburn, NH, 2023).

Newton's Apple is Now The Fermion (Pingree Hill Publishing, Auburn, NH, 2023).

Cosmos Theory: The Sub-Particle Gambol Model (Pingree Hill Publishing, Auburn, NH, 2023).

Cosmos-Universe-Particle-Gambol Theory (Pingree Hill Publishing, Auburn, NH, 2024).

Fractal Cosmos Curve: Tensor-based Cosmos Theory (Pingree Hill Publishing, Auburn, NH, 2024).

The Eternal Form of Cosmos Theory Third Edition (Pingree Hill Publishing, Auburn, NH, 2024).

Fundamental Constants of Cosmos Theory and The Standard Model (Pingree Hill Publishing, Auburn, NH, 2024).

Geometric Cosmos Geometric Universe (Pingree Hill Publishing, Auburn, NH, 2024).

Available on Amazon.com, bn.com Amazon.co.uk and other international web sites as well as at better bookstores.

CONTENTS

<h1 style="text-align:center">FIGURES and TABLES</h1>

Figure 12.1. Sequence 2 fractionation of a particle into 32 gambols. Each gambol is a copy of $(2^5)^{-1}$ times the mass and structure (but not the spin or internal quantum numbers of the particle. The gambol acquires these aspects by inheritance from the

0. Introduction

This book unravels known experimental data to uncover the hidden features within particles – their masses, their interaction coupling constants and their structure. Elementary particles are reduced to their essential form as "mass-energy aggregations" within miniscule universe shells. Fundamental particles and universes are shown to have similar forms.

Conventional theories of elementary particles use Perturbation Theory within the framework of Quantum Field Theory to determine effective coupling constants and particle masses as well as determining scattering amplitudes in elementary particle interactions. The success of this approach varies. It calculates some quantities such as in pure Quantum Electrodynamics with great accuracy. The status is not well understood in other areas. Renormalized coupling constants and particle masses are difficult to ascertain. The Higgs Mechanism does provide estimates of some quantities. Its success is problematic since it fits input values to obtain results consistent with experiment. The origin of these input values is usually not specified. The result is not a deepening of our understanding of particle physics. See chapters 9, 9-A and 11 for details.

The author's approach in this book, and in previous books, is to take the known experimental values, which are the "renormalized" values that perturbation theory should generate (if the perturbation theory calculations could be performed), and from these renormalized values develop an understanding of the coupling constants, the particle masses, and the interior dynamics of particles. The result is a true deepening of our understanding of elementary particle physics.

Cosmos Theory is "universe friendly." It supports perturbation theory in any number of dimensions (for all universes) using the author's Two Tier Theory to obtain finite results in any order of perturbation theory. (No infinities!) It also supports Connection groups in all universes that "connect" all particles. No hidden particles.

The book proves all fermions and hadrons have a gambol interior of mass-energy of the form described by the author previously.

There are important new details on many topics:

On the Cosmos Parent universe consistency condition,
On coupling constants including Gravitation's G,
On the structures of fermion masses as two sequences,
On the reduction of each sequence to a nested set of "boxes",
On Vector and Scalar bosons and their separation into Higgsian and
          non-Higgsian sectors,
On the spectrum of universes in any dimension,
On the structure of fundamental quarks and leptons as miniscule universe shells,
On the mass-energy of a fermion universe shell,

On particle mass-energy as the mass-energy within shells,
On the total mass-energy spectrum of universes within universes,
On a proposed set of new heavy neutrinos,
On the derivation of Gambol Theory from the mass sequences of fermions,

This book is based on the book, *Geometric Cosmos Geometric Universe*, which derives Cosmos Theory Spaces from Totally Antisymmetric Tensors, Fermion Quantum Field Theory, and Fractals. It is also based on books such as *UltraUnification and the Generation of the Cosmos* that explicitly show the inherent Grand Unified Theory (GUT) in each universe individually and the entire Cosmos.

The basic forms of Cosmos Theory and universes are now understood.

The experimentally obtained constants, which usually have error bars, that are considered in this book are well approximated by expressions involving 2, e, and π. The goal of the procedure followed here is to determine the overall structure of universes and particles.

Realizing that these constants are the values (or their close approximations) that might have been found if precise Quantum Field Theory perturbation theory calculations could be performed, we see the work here is a successful "post perturbation theory" description of universe and particle features.

1. Consistency Condition and Coupling Constants

This chapter presents the coupling constants found in our earlier books such as Blaha (2024g). In subsequent chapters we develop a formulation of these fundamental quantities that appear in Cosmology and Elementary Particle Physics. Together we find a fairly complete formulation of fundamental Physics when account is taken of the structure of Quantum Lagrangian Field Theory and its basis in Yang-Mills symmetries.

1.1 Elementary Particle Constants

The major constants of Elementary Particle Physics are the values of Coupling Constants and the values of fundamental Fermion masses. (The presentation here embodies the consequences of Higgs Models. It also embodies the effects of the renormalization of coupling constants and fermion masses.)

Thus these constants are the quantities that one measures in experiments. The fact that the values found are so close to experimental results indicates that we are closer to an ultimate Physical Reality beyond Perturbation Theory.

1.2 Consistency Condition for the Cosmos Parent Universe

The coupling constant values that we have found resulted from a study of a consistency condition for the existence of a Physical Parent universe that is the ancestor of all child universes. (There may be several Parent universes with each heading a hierarchy of child universes.)

The rationale for the consistency condition is:[1]

The Origin of the Cosmos

We initially chose an $r = 18$ space as the space of the original Parent universe raises several issues. (We now prove it.)

The logic of the origin point: There can be no time before the definition of the Parent space dimension array since no space or universe existed before its definition. Yet a Parent universe cannot exist without the definition of the Parent's dimension array. Thus the dimension array and Parent universe must be simultaneously defined and created.

The only way that one may reasonably resolve this state of affairs is to define both as the consequence of a consistency condition. We suggest the consistency condition for the minimum Parent universe dimension is the requirement that its virtual thermodynamic outward pressure equals the Casimir vacuum inward pressure from its virtual external vacuum Casimir pressure at the point of origin of the virtual universe.[2] Below this minimum dimension the Parent universe cannot exist since the virtual vacuum pressure would cause it to be contracted into non-existence.

The dimension r of the Parent universe fixes the Parent universe Physical Cosmos space dimension as the least dimension whose universe is not contracting. Within the Parent universe there are child universes of lower dimension corresponding to lower dimension Cosmos spaces.

---

[1] Blaha (2024d, (2024e) and (2024f).
[2] Much of this chapter is based on Blaha (2022d) and (2023e).

The consistency condition is based on the equality of the inner pressure of a universe and the external Casimir force. Universes where they are equal (dimension r = 18) mark the boundary between unphysical universes that collapse to a point and possible Parent universes that may expand and thus have child universes. (Child universes where r < 18 are Physical.) Possible Parent universes where r < 18 collapse to a point and are unphysical. *Thus the set of Physical Cosmos spaces have r ≤ 18 and are ten in number.* The consistency condition is

Pressure   **Casimir Force**

$$N\,\Gamma((r-1)/2 + 1)kT/(\pi^{(r-1)/2}a^{(r-1)}) = \pi^{(r-1)/2}\,a^{-r}/[r\Gamma((r-1)/2 + 1)] \qquad (1.1)$$

where Cosmos Theory specifies the number of fundamental fermion species N for a universe of dimension r is $N = 2^{r+4}$.

Using Stirling's approximation and after some algebra we find eq 1.1 becomes

Pressure Measure   **Casimir Force Measure**

$$N \qquad\qquad = \qquad\qquad (2e\pi/r)^r/(\pi a r k T) \qquad\qquad (1.2)$$

for equilibrium. If the left side of the equation were larger then universe, then expansion results. If the left side of the equation were smaller then universe, then contraction to a point results.

Why Ten Physical HyperCosmos Spaces?

An examination of eq. 1.2 shows that there is a critical point in the values of N due to the factor

$$(2e\pi/r)^r \qquad\qquad (1.3)$$

where the number $2e\pi/r$ is a transition point from greater than one to less than one at $r = 2e\pi = 17.08$. At this value of r and thus below the even integer dimension value r = 18 the Casimir force is greater than the thermodynamic pressure corresponding to a change between contraction and expansion of the Parent universe. Thus r = 18 marks the minimally acceptable lowest dimension Parent universe since even integer r < 18 would give a universe that contracts to a point and thus could not have child universes.[3] We choose the Physical Parent space dimension r = 18 as the minimum Physically acceptable even integer dimension. (Child universes have dimensions less than 18.)

It is also important to note that $N = 2^{r+4}$ rises with the power r + 4. Thus the Casimir force factor $(2e\pi/r)^r$ should also rise with the power of increasing r in the consistency condition of eqs. 1.1 and 1.2. Therefore $2e\pi/r$ must be greater than one for consistency – thus requiring the minimum Parent universe dimension r to be $[2e\pi] = 18$ where [] signifies "least integer greater than."

Eq. 1.2 has been evaluated. The consistency condition[4] dimension gives r = 18 if

$$akT = 0.0785m_0/m = 2.98 \times 10^{-8} = 2^{-24.8} \cong 2^{-25} \qquad (1.4)$$

where the radius a is set equal to 1/m where m is the r = 18 Parent universe mass, and where m_0 is the gambol mass.[5] The gambol mass is related to the

---

[3] Child universes are not subject to the consistency condition and may expand as ours does.
[4] Blaha (2024e).

universe gambol temperature T by $kT = c_g m_0$ with $c_g = 0.0785$. For $r = 18$ the consistency condition requires $m = 2.6 \times 10^6 m_0$ thus leading to eq. 1.4.

The approximate value of akT in eq. 1.4, when combined with the value of N causes the consistency condition in eq. 1.2 to become

$$\pi N 2^{-25} = \pi 2^{-3} = 0.39 = (2e\pi/18)^{18} \tag{1.5}$$

1.2.1 U(0) Interpretation of akT

The numeric value of akT in eq. 1.4 may be given a physical interpretation based on the coupling constant α_0 of the unitary symmetry group U(0) described in section 1.3 and used several times later in this book in the numerics of universes.

We begin by defining the gambol mass m_0 with[6]

$$m_0 = e^{-4}\alpha_0^{\,2}m = 5.96 \times 10^{-8}m \tag{1.6}$$

where α_0 appears in table 1.3 below. Then

$$akT = 0.0785m_0/m = 4.68 \times 10^{-9} = 2^{-27.67} \tag{1.7}$$

yielding equality for $r - 17.411$:

$$\pi r 2^{r + 4 - 27.67} = 0.714 \tag{1.8}$$

while

$$(2e\pi/17)^{17} = 0.714 \tag{1.9}$$

Since the dimension r must be an even integer, we choose $r = 18$ as the dimension of the largest Physical Cosmos space's universe. This $r = 18$ space's universe is an expanding universe. It is the minimal even dimension space for an expanding universe. Its universe does not contract to a point. (The $r = 17$ possible Parent universe contracts to a point.)

Considering the magnitudes of the masses, eqs 1.8 and 1.9 are an interesting match. This use of the coupling of α_0, which might be viewed as accidental, becomes significant due to four additional repeated appearances in other calculations involving universes that we will see later. *Universes, including Parent universes, appear to have a "duplex" nature with two masses: a "conventional" mass and a gambol mass based on the α_0 coupling constant.*

This use of α_0 in the study of spaces and universes is the first of five appearances in this book. It plays an important role in universe properties.

---

[5] See Blaha (2023e) which shows the appearance of gambol models in many particle and Cosmological cases and the importance of the gambol temperature equation.

[6] The constant e is the base of natural logarithms: $e = 2.718 \ldots$

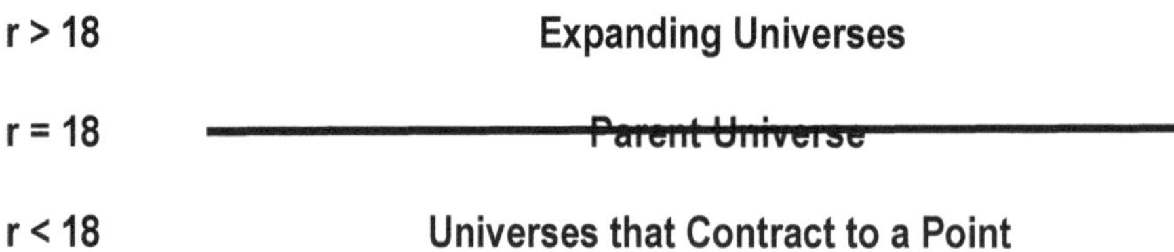

Figure 1.1 The Physical Parent Cosmos universe implied by the consistency condition. The dimension r specifies the space of the universe. It also is the space-time dimension of the universe that is allocated from within the universe space's dimension array.

The key dimension scale for universe and thus space dimension is 2eπ. We view this result as a confirmation of the approach in this chapter and the choice of r = 18 as the minimal possible Physical HyperCosmos Parent space in the 10 Physical HyperCosmos spaces spectrum.

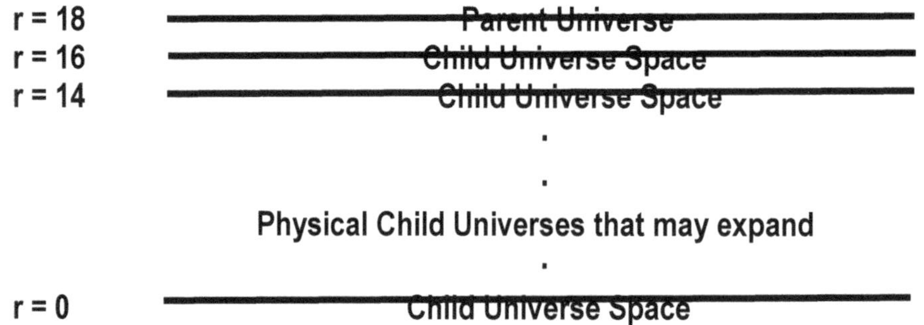

Figure 1.2 The 10 HyperCosmos "physical" spaces that support child universes. The dimension r specifies the space of a universe. It *also* is the space-time dimension of the universe that is allocated from within the universe space's dimension array.

1.2.2 Estimate of Parent Universe Mass-Energy

Given the value for the mass m_0 in eq. 1.6 we can find the mass m of the Parent universe

$$m = (e^{-4}\alpha_0^{2})^{-1}m_0 = (5.96 \times 10^{-8})^{-1}m_0$$

A reasonable choice for the mass m_0 is the estimate of the observable total mass-energy of our universe 3.49×10^{82} GeV/c^2 giving a Parent universe mass-energy:

$$m = 5.86 \times 10^{89} \text{ GeV/c}^2$$

sufficient for more than 16.million universes of the size of our universe.

1.3 Coupling Constant Values

An examination of the form of eq. 1.3 suggests that the dimension $2 \cong e\pi/4$. This suggestion led the author to find a set of almost precise values of symmetry group coupling constants based on e, π and 2. Fig. 1.3 below shows an extremely regular pattern in the coupling constants of the known symmetry groups as well as U(0) – discussed later, and U(256) which appears relevant for the understanding of the gravitation constant G.

The growth in g values by factors of 2 appears to have first been noted by this author in Blaha (2018e) and (2019a) (page 58). This growth supports the quite accurate expression of values in powers of 2. Note that the values are not "bare" values but rather *renormalized values* up to small deviations comparable to experimental uncertainties.

	E S T I M A T E			E X P E R I M E N T		
Interaction	**Expression**	**$g^2/4\pi$ Value**	**g**	**Known[7] Value $g^2/4\pi$**		**Deviation**
U(0) α_0	$e^2/4096$	0.0018036	-	-		-
U(1) $\alpha_1 = \alpha$	$e^2/1024$	0.00721438	0.303	0.0072973525643		1.15%
SU(2) $\alpha_2 = g^2/4\pi$	$e^2/256$	0.0289	0.63	0.0316		9.3%
SU(3) $\alpha_3 = \alpha_S$	$e^2/64$	0.115	1.21	0.117		1.7%
SU(4)[8] α_4	$e^2/16$	0.462	2.4?	0.458		0.087%

$$\cdot$$
$$\alpha_n = g_n^2/4\pi = e^2\, 2^{2n-12}$$
$$\cdot$$
$$\cdot$$

SU(256) α_{256}	$e^2 2^{500}$	2.418×10^{151}	1.74×10^{76}	-		-

Figure 1.3. Coupling Constants table from Blaha (2024h).

The coupling constants have the power representation

$$\alpha_n = g_n^2/4\pi = e^2\, 2^{2n-12}$$
$$= \alpha_0\, 2^{2n} \qquad (1.10)$$

where

$$\alpha_0 = e^2/4096 = e^2 2^{-12}$$

The coupling constants table may be extended backwards to fractional values of n for unitary groups of the form of U(1/n). These groups, which we have considered in earlier books on gambols, have the coupling constants:

$$\alpha_{1/n} = g_{1/n}^2/4\pi = e^2\, 2^{2/n-12} \qquad (1.11)$$

---

[7] All coupling constant values are based on data from Particle Data Group Tables of 2024.
[8] This value is based on the "doubling trend" seen in the three known coupling constants g above.

For example the unitary group U(¼) has the coupling constant

$$\alpha_{1/4} = g_{1/4}{}^2/4\pi = e^2 2^{2/4-12} = e^2\, 2^{-11.5}$$

1.4 Role of the Unitary U(0) Group

  The U(0) group has two elements $\pm I$. It might appear in Lagrangian "mass-like" terms such as

$$g_0\, \overline{\psi}_i U(0) m_{ij} \psi_j \tag{1.12}$$

where $i > j$ and $i, j = \nu_e, \nu_\mu, \nu_\tau, \nu_{\tau'}$.

1.5 Inclusion of Gravitational Coupling Constant G

  The gravitational coupling constant G of our universe can be put in the coupling constant framework that we have established.

  We view the G coupling constant as embodying SU(256):

$$\alpha_{256} = g^2/4\pi = e^2 2^{500} \tag{1.13}$$

We now find the mass associated with gravitational coupling:

$$G = g_{256}{}^2/4\pi\, M^{-2} = 0.67 \times 10^{-38}\ (\text{GeV}/c^2)^{-2} \tag{1.14}$$

We set

$$g_{256}{}^2/4\pi = \alpha_{256}$$

since the 256 fundamental fermions (both normal and Dark) in UST correspond to a SU(256) symmetry group. Then

$$M^2 = (g_{256}{}^2/4\pi)/G = 2.418 \times 10^{151}/(0.67 \times 10^{-38}) = 3.61 \times 10^{189}\ (\text{GeV}/c^2)^2$$

and

$$M = 6 \times 10^{94}\ \text{GeV}/c^2 = g^{-2}\, M_{observable} = 3.5 \times 10^{13}\, M_{observable} \tag{1.15}$$

The value of $M_{observable}$ is a factor of 3.5×10^{13} times the estimated total[9] mass of the *visible* universe:[10]

$$M_{observable} = 1.71 \times 10^{81}\ \text{GeV}/c^2 = 2^{269.85}\ \text{GeV}/c^2$$

  We now note the discrepancy between M and $M_{observable}$ has a simple representation in terms of α_0 and e. Setting

$$M_{observable} = g^2 M \tag{1.16}$$

we find

---

[9] This value is for normal masses - Not mass-energy..

[10] There are many suggestions of a universe mass-energy much greater than the observable.

$$g^2 \cong e^{-6}\alpha_0^{\;4} = e^2(4096)^{-4} = (3.81 \times 10^{13})^{-1} = (2.62 \times 10^{-14})^{-1} \qquad (1.17)$$

with a small deviation of 1.089 or 8.9% in the ratio $3.81 \times 10^{13}/(3.5 \times 10^{13})$.

This use of α_0 in the study of spaces and universes is the second of five appearances in this book. It plays an important role in universe properties. We find the mass of the observable (visible) universe is proportional to the total mass of the universe based on our calculation of G.

This result suggests that the universe is a form of particle with mass M and symmetry group SU(256). It also suggests the universe is much larger than the observable (visible) part. We have suggested this possibility several times previously in other contexts.

SU(256) like SU(3) might have been associated with confinement. However this possibility is removed by the mass factor M^{-2} that is required on dimensional analysis grounds in gravitational dynamic equations.[11] This factor causes the gravitation G to be weak.

We appear to have a viable estimate of the value of G and its origin in a coupling constant formulation.

Later we extend this discussion to the case of all universes.

1.6 Derivation of the Form of Coupling Constants

We have shown coupling constants appear to form a powers sequence. We now consider the expression of a coupling constant as the product of a number of factors that are based on geometry.

First we note that n is the number of fermions in a fundamental representation of SU(n). The coupling constant $g(n) = g_n$ has the form:

$$g(n) = e\,(4\pi)^{\frac{1}{2}}\,2^{n-6} = e\,(2^{-10}\pi)^{\frac{1}{2}}\,2^n = 0.1505\;2^n \qquad (1.18)$$
$$= g_g\,(\text{number of spin states per fermion})^{\text{number of fermions}}$$

The coupling constant is a fundamental value g_g times the number of spin states (a geometric property of fermions equal to half the column length of a spinor) raised to the number of fermions. The number of spin states per fermion is 2 in $r = 4$ dimensions.

We see that $g(n)$ has n factors – one factor for 2 spin states for each fermion in the SU(n) fundamental representation. The origin of

$$g_g = e\,(2^{-10}\pi)^{\frac{1}{2}} \qquad (1.19)$$

Also

$$\alpha_n = g_n^{\;2}/4\pi = e^2\,2^{2n-12}$$
$$= \alpha_0\,2^{2n} \qquad (1.20)$$

---

[11] Models which change gravity from 1/r Newtonian gravity usually change the space-time dependence – not G.

Eq. 1.18 is *initially* not decisively understood due to the appearance of the natural logarithm base e. However e and π appear naturally later in our studies of higher dimension volumes and surface areas. We therefore view g_g as having a *geometric origin.*

Eq. 1.18 implies that the coupling g(n) for SU(n) is g_g times n factors of 2 which represents the number of spin states per fermion in the SU(n) fundamental representation in four dimensions. Thus it is a product of degrees of freedom.

An important point is the number of factors is equal to the number of fermions in the SU(n) fundamental representation. *The coupling constant "felt" by one fermion depends on the total number of fermions in the group's fundamental representation.*

Here again we find a dependence on α_0 favoring the "duplex" interpretation.

2. Fermion Masses

2.1 Fermion Generations

The Standard Model *defines* fermion generations (Fig. 2.1). Cosmos Theory and, in particular, The Unified SuperStandard Theory (UST) within it, define generations in a manner based on the form of the r = 4 dimension array of the Cosmos Spectrum.[12] The UST has four layers, which each contain four generations defined differently from The Standard Model.[13]

Each UST generation has eight "normal" fundamental fermions. (There is also a corresponding eight Dark fundamental fermions in the Dark matter sector.) Thus there is a total of 4×4×8 = 128 normal fermions and 128 Dark fermions.

The UST dimension array is square 16×16 (Half of the array is Dark sector.) as are all HyperCosmos spaces dimension arrays. This is shown in our Fractal books on Cosmos Theory, which show their analogue γ-matrices are also square. Both types of arrays follow from the form of totally antisymmetric tensors.

The Standard Model Generations

Generation:	1	2	3
	u	c	t
	d	s	b
	e	μ	τ
	ν_e	ν_μ	ν_τ

Figure 2.1. The Standard Model generations.

We will consider the *Current* mass spectrum of the *eight* first generation "normal" fermions of UST. See Figs. 2.2 and 2.3. We will show they exhibit interesting regularities, which support the UST (and Cosmos) view of generations and also provide a simple, surprising, numerical fermion spectrum of current masses.

2.2 Comparison Between Standard Model and UST Generations

The Standard Model may be formulated without generation quantum numbers. The generations diagram of Fig. 2.1 is not required by the equations and calculations of The Standard Model since they may be (and are) usually written without "generation quantum numbers" attached to states in perturbation theory.

---

[12] See Blaha (2023d) and (2018e) for details.

[13] The four layers support the author's Layer groups and the four generations per layer support the author's Generation groups. There are also Connection groups. See Appendices 25-A and 25-B of Blaha (2024h) for a description of these groups, all of which were originally introduced some time ago by the author.

On the other hand The Unified SuperStandard Theory (UST) has Generation groups – one U(4) Generation Group for each of the four layers of fermions in Fig. 2.2. In each generation, in each layer, a generation quantum number is assigned to each fermion. Thus, for example, all fermions in generation 1 in layer 1 may be assigned the generation number 1.[14] All fermions in generation 2 in layer 1 may be assigned the generation number 2 and so on. Similarly, the second layer which has its own U(4) Generation group may have quantum numbers assigned to the fermions of each of its generations. The four U(4) Generation Groups of the four layers are separate and have their own sets of generation quantum numbers for the fermions in each layer.

Thus the UST has a well-defined pattern of generations in layers with generation group quantum numbers. Similar comments apply to the Layer groups. Each generation has a separate Layer group. Thus there are four U(4) Layer groups – one group for each of the four generations spread over the four layers. Thus the first generation in each layer may be assigned U(4) Layer group quantum numbers 1, 2, 3, 4 for Layer group 1 The first generation in layer 1 has layer number 1. The first generation in layer 2 has layer number 2 and so on..

Similarly, the second generation in each of the four layers may be assigned layer group quantum numbers 1, 2, 3, 4. And so on for Layer group 3 and Layer group 4.

The Generation and Layer groups[15] are discussed in detail in Appendix 25-A in Blaha (2024h) and in earlier books such as Blaha (2018e). See Fig. 2.3 for lines connecting elements of the fundamental representations of Generation groups and Layer groups.

*We conclude the UST has the correct generations and layers **for the UST**. UST calculations for the first generation of the first UST layer are the same as The Standard Model equivalents. Fig. 2.1 is acceptable for the more limited Standard Model.*

2.3 Experimentally Known Fermion Masses

At first glance the first generation mass spectrum of Fig. 2.4 below does not appear to exhibit any regularities. In previous books we developed a new view of the spectrum based on the diagram in Fig. 2.5.

This diagram portrays the fermion masses on two levels based on the ElectroWeak separation of fermions in pairs. On each level the masses are ordered in increasing value. The diagram inverts the u and d level locations based on the overall ordering of top level masses as above the corresponding low level masses. Each ElectroWeak pair has the more massive above the less massive.

---

[14] This generation consists of the eight fundamental fermions of Fig. 2.2.

[15] The Generation and Layer groups only make sense in the UST dimension array form. That is why the author introduced them originally. They are not reasonable in the conventional Standard Model and thus were not introduced in that context leaving it to the author to first introduce them within the UST. See Blaha (2018e).

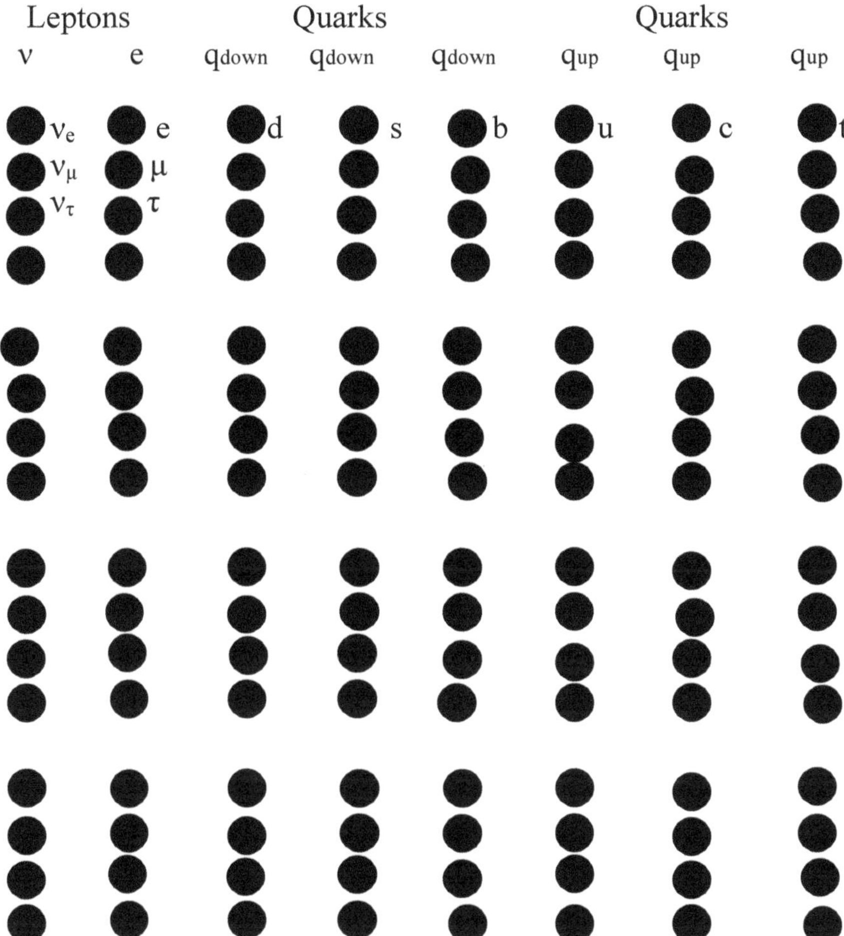

Figure 2.2. Four UST (and Cosmos) layers with each consisting of four generations of eight normal and eight Dark fermions. From Blaha (2023d) and the author's earlier books.

The UST Fermion Periodic Table

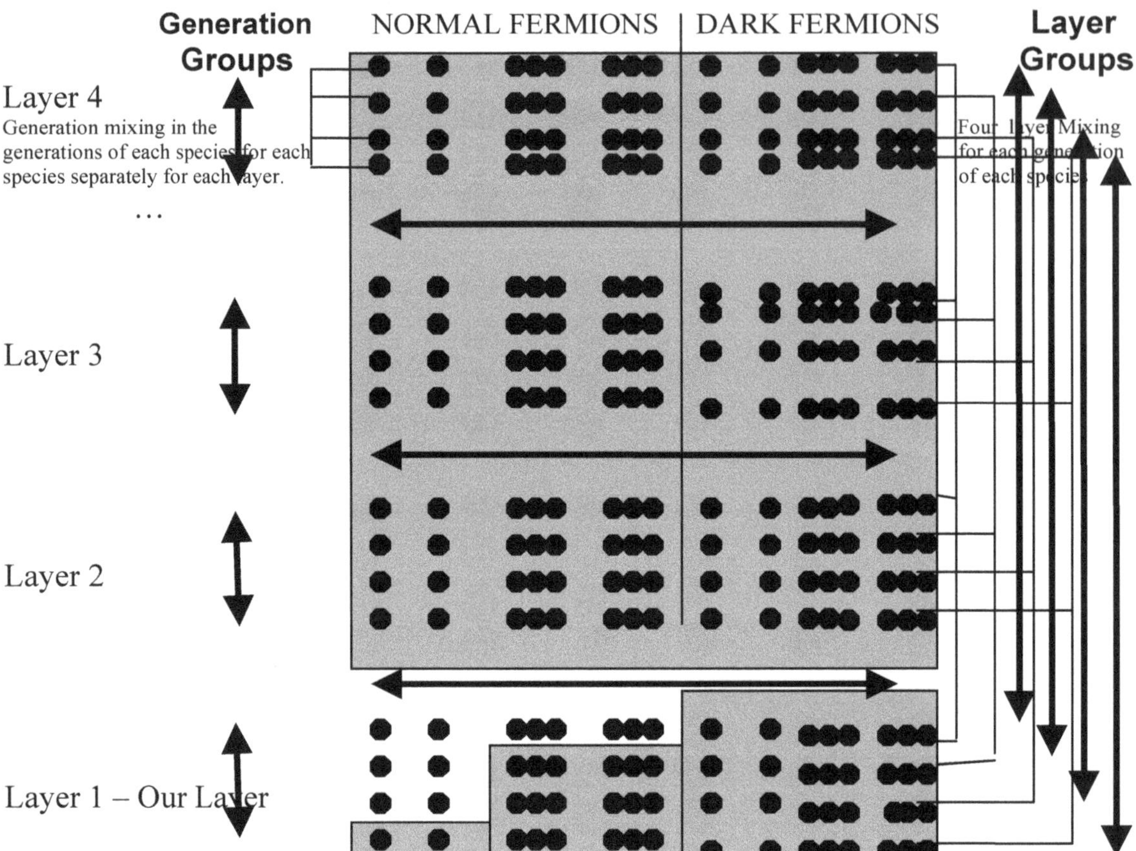

Figure 2.3. from Fig. 6.4 of (2024h). Fermion particle spectrum and partial example of pattern of mass mixing of the Generation and Layer grroups. Unshaded parts are the known fermions including an additional, as yet not found, 4th generation shown. The lines on the left side (only shown for one layer) display the Generation mixing within each layer's species. The Generation mixing applies within each layer using a separate Generation group for each layer. The lines on the right side show Layer group mixing with the mixing amongst all four layers for each of the four generations individually. There are four Layer groups. For each generation and each layer, SU(2)⊗U(1) mixes between an e-type fermion and a neutrino-type fermion. It also mixes between an up-quark-type fermion and a down-quark-type fermion. SU(3) mixes among each up-quark triplet and down-quark triplet separately. There are 256 fundamental fermions counting quarks as triplets. This figure is from Blaha (2020e) for the UST.

$v_e = 3.72 \times 10^{-11}$ GeV/c^2

> Our v_e mass estimate $= .03878$ ev/c^2 is consistent with known data estimates.

$v' = e^2(\alpha_0{}^2)^{-1}v_e = 0.0372$ ev/c$^2 \times 4096^2/e^2 = 8.4 \times 10^{-5}$ GeV/c^2 where $e = 2.718$ and $\alpha_0 = e^2/4096$.

> We scale v_e by $e^2(\alpha_0{}^2)^{-1}$.to obtain a proposed v' neutrino. This NEW neutrino has a mass consistent with the scale of other first generation fermion masses. We will use this particle and its mass to study fermion mass regularities. Note: the use of $e^2(\alpha_0{}^2)^{-1}$. This again reflects a further structural regularity that we call duplex.

$e \;= 0.511 \times 10^{-3}$ GeV/c^2 = electron mass

$u \;$ 1.76 mev/c2 $= 1.76 \times 10^{-3}$ GeV/c^2

> **We scale the u mass by a factor of 4/5** – again to put it on a similar footing as the other masses.
> Note: the u mass is 2.2 +0.5 -0.4 $\times 10^{-3}$ GeV/c^2 The 1.76 value is consistent with the lower bound 1.8.

$c \;$ 1.27 GeV/c^2

$t \;$ 172.76 GeV/c^2

$d \;$ 3.76×10^{-3} GeV/c^2

> **We scale the d mass by a factor of 4/5** – again to put it on a similar footing as the other masses.
> Note: the d mass is 4.7 +.5 - .3 $\times 10^{-3}$ GeV/c^2 The 3.76 value is consistent with the lower bound 4.4.

$s \;$ 95×10^{-3} GeV/c^2

$b \;$ 4.18 GeV/c^2

Figure 2.4 The **experimentally known** Fundamental Fermions from Fig. 16.4 of Blaha (2024g). Fundamental first generation current fermion masses. **Each fundamental fermion symbol represents its mass.** We have augmented the list with a proposed v' neutrino that seems appropriate given the masses of the other fermions. We will discuss it later.

We introduce vertical arrows in Fig. 2.5 pointing to the lower mass of corresponding pairs. We introduce "angled" arrows between consecutive pairs pointing to the larger mass of adjacent pairs. This procedure generates a path between masses that leads to surprising regularities in fermion mass relations.

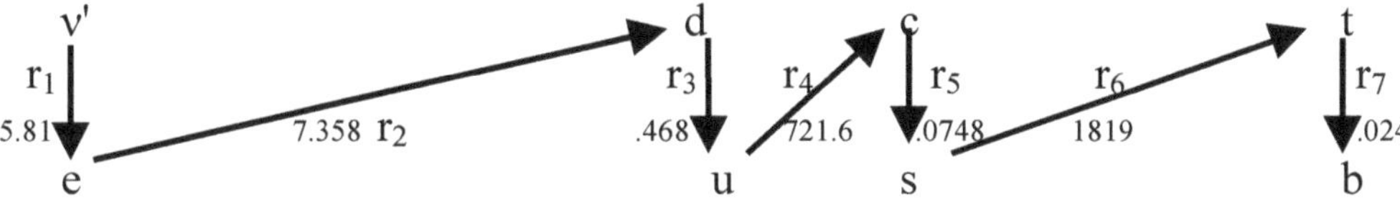

Figure 2.5. Fundamental first generation current fermion masses. Each fundamental fermion symbol represents its mass. Note the charges and internal quantum numbers differ for fermions connected by arrows. The focus is on mass relations following from internal dynamics modulo quantum numbers.

Fig. 2.5 displays ratios denoted r_i where i =1, 2, ... , 7 that we will now define and use to generate relationships between fermion masses. We define the ratios with their corresponding numeric values obtained directly from Fig. 2.4.

$$r_1 = e/\,v' = .511\times 10^{-3}/(8.4\times 10^{-5}) = 6.08 \tag{2.1}$$
$$r_2 = d/e = 7.358$$
$$r_3 = u/d = 0.468$$
$$r_4 = c/u = 721.6$$
$$r_5 = s/c = 74.8\times 10^{-3} = 0.0748$$
$$r_6 = t/s = 1.819\times 10^{3} = 1819$$
$$r_7 = b/t = 0.024$$

We can use these ratios and values to define fermion masses. Note first

$$v' = e^{2}v_e\,(\alpha_0{}^{2})^{-1} \tag{2.2}$$
$$r_2/r_1 = 1.21 \cong 4/\pi = 1.27$$

Then we use eq. 2.1 to obtain

$$e = r_1\,v' \tag{2.3}$$
$$d = r_2e = r_2r_1v'$$
$$u = r_3d = r_3r_2r_1v'$$
$$c = r_4u = r_4r_3r_2r_1v'$$
$$s = r_5c = r_5r_4r_3r_2r_1v'$$
$$t = r_6s = r_6r_5r_4r_3r_2r_1v'$$
$$b = r_7t = r_7r_6r_5r_4r_3r_2r_1v'$$

This buildup is in "slice" factors of *mass* in a manner analogous to the buildup of coupling constants with spin degrees of freedom factors in section 4.2 of Blaha (2024f).

Following the arrows in Fig. 2.4 we see two patterns emerge. The set of angled arrows exhibit an increasing sequence of ratios. We form ratios of ratios and see approximate factors containing π appear:

$$
\begin{array}{llll}
r_2 & r_4 & r_6 & \qquad(2.4)\\
7.358 & 721.6 & 1819 \\
\quad r_4/r_2 = 98.1 & & r_6/r_4 = 2.52 \\
\quad\quad \cong 32\pi & & \quad \cong \pi
\end{array}
$$

The set of vertical arrows exhibit a decreasing sequence of ratios:

$$r_1 \qquad\qquad r_3 \qquad\qquad\qquad r_5 \qquad\qquad r_7 \qquad\qquad\qquad (2.5)$$

$$6.08 \qquad\qquad 0.468 \qquad\qquad\qquad 0.0748 \qquad\qquad 0.024$$

$$r_1/r_3 = 12.99 \qquad r_3/r_5 = 6.28 \qquad\qquad r_5/r_7 = 3.116$$

$$\cong 4\pi \qquad\qquad\qquad \cong 2\pi \qquad\qquad\qquad\qquad \cong \pi$$

The angled line data gives:

$$r_4 = 32\pi r_2 \qquad\qquad r_6 = \pi r_4 = 32\pi^2 r_2 \qquad\qquad (2.6)$$

Note the powers of $\pi^{(i-2)/2}$ for i = 4, 6.
The vertical line data gives:

$$r_1 = 6.08$$

$$r_5 = \pi\, r_7 \qquad\qquad r_3 = 2\pi r_5 \qquad\qquad r_1 = 4\pi r_3 \qquad\qquad (2.7)$$

Using the above relations we find

$r_1 = e/\,v' = 6.08 \cong 2\pi \qquad\qquad\qquad\qquad\qquad\qquad\qquad\qquad (2.8)$

$r_2 \cong 4/\pi\; r_1 \cong 8$

$r_3 \cong r_1/(4\pi) \cong 1/2$

$r_4 \cong 256\pi$

$r_5 \cong 1/(4\pi)$

$r_6 \cong 256\pi^2$

$r_7 \cong 1/(4\pi^2)$

with $v' = 8.4 \times 10^{-5}$ GeV/c^2. Fig. 2.5 shows the resulting approximate mass values.

INITIAL ESTIMATE	**DEVIATION RATIO**	**DATA**
$e = r_1 v' = 2\pi v' = 0.528 \times 10^{-3}$	1.08	0.511×10^{-3} GeV/c^2
$d = r_2 e = r_2 r_1 v' = 2^4\pi v' = 4.42 \times 10^{-3}$	1.18	3.76×10^{-3} GeV/c^2
$u = r_3 d = r_3 r_2 r_1 v' = 2^3\pi v' = 2.51 \times 10^{-3}$	1.42	1.76×10^{-3} GeV/c^2
$c = r_4 u = r_4 r_3 r_2 r_1 v' = 2^{11}\pi^2 v' = 1.78$	1.4	1.27 GeV/c^2
$s = r_5 c = r_5 r_4 r_3 r_2 r_1 v' = 2^9\pi v' = 141.55 \times 10^{-3}$	1.49	95×10^{-3} GeV/c^2
$t = r_6 s = r_6 r_5 r_4 r_3 r_2 r_1 v' = 2^{17}\pi^3 v' = 357.64$	2.07	172.76 GeV/c^2
$b = r_7 t = r_7 r_6 r_5 r_4 r_3 r_2 r_1 v' = 2^{15}\pi v' = 9.06$	2.16	4.18 GeV/c^2

Figure 2.6. Approximate estimates of the fermion current masses. The estimates can be viewed as approximations to the experimental data based on powers of 2 and π.

The agreement of the estimates[16] formed from directly examining the data, and the actual data, is good for e and d; is less good for u, c, and s; and much less good for t and b. We remedy these anomalies next.

2.4 Revised Estimate of Current Fermion Masses based on SU(2)/SU(3) Variation

In the preceding section we developed the forms of eight known fundamental fermion masses based on a constant times a power of 2 and a power of π. These forms are analogous to the forms of coupling constants that we recently found based on powers of 2 and e – the base of the natural logarithms. Together this harmony of mass and coupling constant forms lends indirect support to Cosmos Theory. It is somewhat remarkable that coupling constants are based on e and masses are based on π. They are not jointly based on both. It suggests there is a deeper, new layer of particle interactions which we will elucidate later.

We now turn to the possibility of improving the mass estimates of section 3.2 by the introduction of a factor giving a sliding scale of mass dependence in the ratio of an ElectroWeak SU(2) factor and a Strong Interaction SU(3) factor. We assume they contribute with equal strength to the e and d masses; with a mild dominance of ElectroWeak over Strong for u, c, and s quarks; and with dominance of Strong over ElectroWeak by a factor of 2 for t and b quarks.

We denote the ratio of the interaction strengths with

$$\alpha/\beta \equiv \alpha_{SU(2)}(m)/\beta_{SU(3)}(m) \qquad (2.9)$$

where m is the mass of the relevant fermion. We note that the ratio of 2 for t and b reflects the ratio of the coupling constants ($1.21/0.63 = 1.92 \cong 2$). The value of α/β nicely progresses from 1 to ¾ to ½ as the fermion mass increases.

α/β	REVISED ESTIMATE	DEVIATION RATIO	DATA
1	$\nu_e = 0.0372 \text{ ev}/c^2 = 3.72 \times 10^{-11} \text{ GeV}/c^2$		
1	$\nu' = \nu_e\, e^2(\alpha_0^2)^{-1} = 8.4 \times 10^{-5} \text{ GeV}/c^2 \cong e2^{-15} \text{ GeV}/c^2$		
1	$e = r_1\, \nu' = 2\,\pi\,\nu' = 0.511 \times 10^{-3}$	1.	$0.511 \times 10^{-3} \text{ GeV}/c^2$
1	$d = r_2 e = r_2 r_1 \nu' = 16\pi\,\nu' = 4.24 \times 10^{-3}$	1.13	$3.76 \times 10^{-3} \text{ GeV}/c^2$
¾	$u = r_3 d = r_3 r_2 r_1 \nu' = ¾ \times 2^3 \pi\,\nu' = 1.80 \times 10^{-3}$	1.02	$1.76 \times 10^{-3} \text{ GeV}/c^2$
¾	$c = r_4 u = r_4 r_3 r_2 r_1 \nu' = ¾ \times 2^{11} \pi^2 \nu' = 1.28$	1.001	$1.27 \text{ GeV}/c^2$
¾	$s = r_5 c = r_5 r_4 r_3 r_2 r_1 \nu' = ¾ \times 2^9 \pi\,\nu' = 102 \times 10^{-3}$	1.07	$95 \times 10^{-3} \text{ GeV}/c^2$
½	$t = r_6 s = r_6 r_5 r_4 r_3 r_2 r_1 \nu' = ½ \times 2^{15} \pi^3 \nu' = 171$	0.99	$172.76 \text{ GeV}/c^2$
½	$b = r_7 t = r_7 r_6 r_5 r_4 r_3 r_2 r_1 \nu' = ½ \times 2^{13} \pi\,\nu' = 4.34$	1.04	$4.18 \text{ GeV}/c^2$

Figure 2.7. Revised fermion masses based on α/β and on the known value of e of 0.511 from which the values of v' and v are extracted and used to calculate the other fermion masses. The d discrepancy is the largest reflecting, perhaps, some unceratinty in the experiment value of d. Otherwise there is excellent agreement with experiment!

---

[16] The limitation to powers of 2 and π times v' makes the estimates non-trivial.

Strong SU(3) dominates at high mass; ElectroWeak SU(2) dominates at low mass. All estimates in Fig. 2.7 are close to the experimental values. Thus we have a representation of fermion masses in terms of powers of 2, powers of π, and the value of v' - comparable to coupling constants where coupling constants are powers of 2 multiplied by e – the base of natural logarithms. This suggests that Cosmos Theory is at the base of both coupling constant and fermion mass values. Why one uses π and e, and the other uses only e - the base of natural logarithms, will be shown to be the result of Geometry later.

Substituting $v' \cong e2^{-15}$ GeV/c^2 gives Fig. 2.8.

α/β	FERMION
1	$v_e = 0.0372$ ev/c$^2 = 3.72 \times 10^{-11}$ GeV/c^2
1	$v_e = \alpha_0^2/e^2$ $v' = e^3 2^{-39}$ GeV/c^2
1	$v' = e^2(\alpha_0^2)^{-1} v_e \cong e\, 2^{-15}$ GeV/c^2
1	$e = r_1 v' = \pi e2^{-14}$ GeV/c^2
1	$d = r_2 e = r_2 r_1 v' = \pi e 2^{-11}$ GeV/c^2
¾	$u = r_3 d = r_3 r_2 r_1 v' = ¾ \times \pi e 2^{-12}$ GeV/c^2
¾	$c = r_4 u = r_4 r_3 r_2 r_1 v' = ¾ \times \pi^2 e 2^{-4}$ GeV/c^2
¾	$s = r_5 c = r_5 r_4 r_3 r_2 r_1 v' = ¾ \times \pi e 2^{-6}$ GeV/c^2
½	$t = r_6 s = r_6 r_5 r_4 r_3 r_2 r_1 v' = 2\pi^3 e$ GeV/c^2
½	$b = r_7 t = r_7 r_6 r_5 r_4 r_3 r_2 r_1 v' = \pi e 2^{-1}$ GeV/c^2

Figure 2.8. Revised form of fermion masses using $v' \cong e2^{-15}$ GeV/c^2.

2.5 Fermion Masses are Analogous to Coupling Constants

A comparison of Figs. 2.8 and 2.9 below shows a significant resemblance which we take to imply that fermion mass is analogous to a coupling constant. Note fermion masses couple the large and small parts of spinors. Note the appearance of e, the base of natural logarithms. Note also the powers of two. Masses differ from coupling constants in having powers of π.

Interaction		$g = (4\pi\, \alpha)^{½}$
U(0)		$g_0 = e2^{-6}$
U(1)	$\alpha_1 = \alpha$	$g_1 = e2^{-5}$
SU(2)	$\alpha_2 = g^2/4\pi$	$g_2 = e2^{-4}$
SU(3)	$\alpha_3 = \alpha_S$	$g_3 = e2^{-3}$
SU(4)	α_4	$g_4 = e2^{-2}$

.

.

.

Figure 2.9. Coupling constants.

2.6 Two Sequences of Fermion Masses

We have found two sequences of fermion masses apparent in Fig. 2.6:

	SEQUENCE 1				SEQUENCE 2		
	e	u	c	t	ν'	d	s	b
Mass:	0.511×10^{-3}	1.80×10^{-3}	1.28	171	8.4×10^{-5}	4.24×10^{-3}	102×10^{-3}	4.34
Multiplier:		$2^5\pi$	$2^5\pi$	$2^5\pi$		32	32	32

where the multiplier connects adjacent masses in each sequence separately.

We suggest all four generations of fermions implement this separation into sequences.

The two sequences implement the multipliers:

Up type fermion sequence multiplier: $2^5\pi = 32\pi$ Sequence 1
Down type fermion sequence multiplier: $2^5 = 32$ Sequence 2

We believe the multiplier of $2^5 = 32$ appearing in both sequences is the number of fermions in a UST layer based on the concept that all fermions of a layer are intertwined. There are 32 normal fermions in a UST layer. The factor of π for sequence 1 multiplies each of the 32 slices' mass in a fermion by a factor of π.

The explanation of the factors of π reflects the internal universe shells, which we will discuss later, is Geometry.

Note each mass has implicit e and π factors due to

$$\nu' = e^2(\alpha_0^2)^{-1}\, \nu_e = 8.4 \times 10^{-5} \cong e\pi \times 10^{-5}\ \text{GeV/c}^2 \qquad (2.10)$$

2.7 Factoring Mass Values in the Manner of Gambol Theory

Coupling constant values depend on powers of 2 with the power of 2 determined by the count of spin degrees of freedom of the fermions in the group's fundamental representation. Based on this consideration it is reasonable to consider factoring mass values in a manner similar to the factoring of particles into gambols. Gambol Theory defines the gambol mass m_g of a particle with a similar factoring equation:

$$m_g = M/s \qquad (2.11)$$

where s is a power of 2 and M is a fermion mass, as we saw in our earlier books.

We now can consider factoring a mass such as the u quark mass into gambol masses. For example we can factor the u quark into eight parts with

$$m_g = \tfrac{3}{4} \times \pi e 2^{-12}/2^8 = = \tfrac{3}{4} \times \pi e 2^{-20} \qquad (2.12)$$

and then proceed to consider applications in the Gambol Theory of quarks.

Factoring the masses of fermions thus leads to a deeper view of fundamental fermions as having a type of conceptual partitioning.

Later we will see how the sequences of fermions lead to an understanding of the interior of fermions and thence to the Gambol Theory.

3. Proposed New Neutrino

In chapter 2 it appeared reasonable to introduce a new heavy neutrino ν' in the first generation and correspondingly in other fermion generations. This neutrino was related to the U(0) unitary group in chapter 1 where it appeared as an extension of the known internal symmetry groups.

The U(0) group has two elements $\pm I$. It has a coupling constant and could appear in Lagrangian "mass-like" fermion terms such as

$$g_0\,\overline{\psi}_i U(0) m_{ij} \psi_j \tag{1.11}$$

where $i > j$ and $i, j = \nu_e, \nu_\mu, \nu_\tau, \nu_{\tau'}$.

On this basis, and noting the significant difference in mass of the neutrinos, which are both of generation 1, one may assume that a ν' should decay into a "normal" neutrino ν rapidly. Thus the only likely appearance of ν' might be in Cosmic ray showers. Unusual Cosmic ray events, such as the Niu event, that might contain ν' neutrinos have been seen many times. Their detection would have major consequences.

At the moment we can only say that a ν' fermion should have the same features (spin and internal symmetries)[17] as a ν_e neutrino but have much larger mass. We estimate the ν' mass to be of the order of $e^2(\alpha_0^2)^{-1}\,\nu_e \cong e\ 2^{-15} = 8.4\times10^{-5}$ GeV/c^2.compared to the ν mass of the order of 0.0372 ev/c^2 $= 3.72\times10^{-11}$ GeV/c^2.– differing by a factor of approximately 10^7.

---

[17] They would require the Weinberg-Salam ElectroWeak Theory to be augmented.

4. Structure of Mass Sequences

4.1 Mass Spectrums of the Sequences

The preceding study of mass sequences yielded mass formulae that are similar in form to the form of interaction coupling constants.

	Sequence 1				Sequence 2		
n:	1	2	3	4	1	2	3	4
Mass (GeV/c²):	$0.511{\times}10^{-3}$	$1.80{\times}10^{-3}$	1.28	171	8.4×10^{-5}	$4.24{\times}10^{-3}$	$102{\times}10^{-3}$	4.34
Multiplier:	$2^5\pi$	$2^5\pi$	$2^5\pi$		$2^5 = 32$	$2^5 = 32$	$2^5 = 32$	

Figure 4.1. The sequences of fermion masses.

In chapter 12 we will see that the sequences of fermion masses are direct evidence (and thus a proof) of the fractionation of fundamental fermions according to Gambol Theory.

Our study of fundamental fermion masses yielded two sequences. In each sequence we saw simple (approximate) expressions for first generation fermion masses:

$$m_{1n} = (\pi 2^5)^n m_{10} \qquad \text{Sequence 1} \qquad (4.1)$$
$$m_{10} = 0.511{\times}10^{-3}/(2^5\pi) \ \text{GeV/c}^2 = 5.08 \times 10^{-6} \ \text{GeV/c}^2$$

for n = 1, 2, 3, 4 and

$$m_{2n} = (2^5)^n m_{20} \qquad \text{Sequence 2} \qquad (4.2)$$
$$m_{20} = 8.4 \times 10^{-5}/2^5 \ \text{GeV/c}^2 = 2.63 \times 10^{-6} \ \text{GeV/c}^2$$

for n = 1, 2, 3, 4. Note the comparable values of m_{10} and m_{20}:

$$m_{10}/m_{20} = 1.93 \cong 2 \qquad (4.3)$$

The differences in the masses in the sequences are primarily due to the differing multipliers.

4.2 Comparison to Coupling Constants

Fig. 4.1 and eqs. 4.1 and 4.2 are analogous to Fig. 1.3 and eq. 1.10 for coupling constants. This parallel suggests a deeper relation between masses and coupling constants that we pursue later.

5. The Masses of Cosmos Universes

In Blaha (2024h) we saw that the Parent universe consistency condition led to the determination of internal symmetry coupling constants. In chapter 2 we considered the masses of fermions and saw significant regularities. We now turn to develop a condition for the spectrum of possible universe mass-energies at the time of their creation in a parent universe.

This chapter develops a condition for the determination of the initial mass-energy of a universe through an interaction or *ab initio*. A universe is viewed as a type of particle. A universe in its initial state is assumed to have dimension r and to reside in a parent universe of dimension r' > r. The contents of the universe initially have a high temperature T_r that causes its contents to consist of interaction-less, massless particles. We treat the contents of a universe as a perfect gas initially satisfying the Ideal Gas Law.

We assume the initial universe at the creation point is bound by a Casimir vacuum energy force from the parent universe that momentarily balances the pressure of its internal mass-energy contents at the creation point.

This scenario gives a consistency condition similar to that in chapter 11 of Blaha (2024h). We assume the universe is initially an r dimension sphere in an r' dimension parent universe.[18] The calculation of the internal pressure is the same as in chapter 11 of Blaha (2024h).

The calculation of the Casimir force assumes an r' dimension space with a dimension array with $N_{r'} = 2^{r'+4}$ fermions. The estimates of the volume and surface area of the sphere are the same as in chapter 11 of Blaha (2024h).

We thus arrive at the modified (eq. 11.18 of Blaha (2024h)) balance of pressure and Casimir force:

$$0 = N_r \, \Gamma((r-1)/2 + 1)kT/(\,\pi^{(r-1)/2}a^{(r-1)}) - N_{r'}\pi^{(r-1)/2}\,a^{-r}/[r\Gamma((r-1)/2 + 1)] \qquad (5.1)$$

implying

$$E_r = N_r \, kT_r = N_{r'} \, \pi^{(r-1)}(ar)^{-1}/[\Gamma((r-1)/2 + 1)^2] \qquad (5.2)$$
$$= N_{r'}\,\pi^{r-1}(r/2e)^r)/(a\pi r^2)$$
$$= N_{r'}\,(2e\pi/r)^r/(a\pi^2 r^2) \qquad (5.3)$$

after using Stirling's approximation for Γ functions (eq. 11.24 of Blaha (2024h)). E_r is the total mass-energy for a universe of dimension r created in a universe of dimension r'.

---

[18] An ultimate parent universe is a dimension 18 universe.

5.1 Universe Particle Mass Spectrum

The universe mass spectrum depends on the dimension array size of the universe, within which it directly resides:

$$E_{r,r'} = 2^{r'+4}(2e\pi/r)^r/(a\pi^2 r^2) \tag{5.4}$$

relates the mass-energy $E_{r,r'}$ to r and r' and the initial radius a of the dimension r universe.

Eq. 5.4 shows a joint sequence of universe energy as a function of r and r'.
For our $r = 4$ universe if it resides in an $r' = 6$ Megaverse we find

$$E_{4,6} = 2^6(e\pi/2)^4/(a\pi^2) = 2155/a$$

suggesting our universe in this case has an initial radius of

$$a = 2155/E_{4,6}$$

Using the *total* mass-energy of the observable (visible) universe $E_{4,6} = 3.49 \times 10^{82}$ GeV/c^2 we find the initial radius[19] (before/simultaneous with the Big Bang)

$$a = 6.17 \times 10^{-80} \, (GeV/c^2)^{-1}$$

Using the relation between a spatial meters and energy

$$m \equiv 0.89 \times 10^{12} \, (GeV/c^2)^{-1}$$

we find the initial radius a of our universe is

$$a = 5.55 \times 10^{-68} \, m \tag{5.5}$$

5.2 Universe Temperature

The product of the initial temperature of a universe and its radius a satisfies

$$akT_r = N_{r'}/N_r \, (2e\pi/r)^r/(\pi^2 r^2) = 2^{r'-r}(2e\pi/r)^r/(\pi^2 r^2)$$

As $a \to 0$ and $kT \to \infty$, the product remains constant.[20]

5.3 Temperature Related to Mass as in the c_g Relation

Eq. 5.4 can be reexpressed as a temperature – mass relation:

$$kT_{r'} = 2^{r'-r}(2e\pi/r)^r/(\pi^2 r^2) \, m \tag{5.6}$$

---

[19] Blaha (2004) has a very small but non-zero radius at the Big Bang point.
[20] This result is similar to those of Big Bang models such as this author's Blaha (2004) model.

by setting $m = a^{-1}$. This equation is analogous to the relation we found previously in the gambol analysis of particles:

$$kT = c_g m \tag{5.7}$$

where[21]

$$c_g = 0.0785 \cong \pi^2/128 \tag{5.8}$$

We define the quantity

$$C_g = 2^{r'-r}(2e\pi/r)^r/(\pi^2 r^2) = \tfrac{1}{4}\,\pi^2\,(e/2)^4 = 8.41 \cong 0.852\pi^2$$

for $r = 4$ and $r' = 6$ and find

$$c_g/C_g = 1/(.852 \times 128) = 1/109 = 0.0092$$

5.4 Temperature – Mass Relation $kT = c_g M$

The relation in eq. 5.6 was based on identifying $a^{-1} = m$. If we instead require

$$kT_{r'} = c_g\, m \tag{5.9}$$

then

$$kT_{r'} = 2^{r'-r}(2e\pi/r)^r/(a\pi^2 r^2) = \tfrac{1}{4}\,\pi^2\,(e/2)^4/a = \pi^2/128\, m$$

for $r = 4$ and $r' = 6$ and so

$$a = 32(e/2)^4/m = 109/m \tag{5.10}$$

Eq. 5.10 implies eq. 5.8 for c_g.

5.5 Expanded Energy Formula

We now reexpress the Energy expression of eq. 5.4 in terms of m using eq. 5.10.

$$\begin{aligned}
E_{r,r'} &= 2^{r'+r+4}\, r^{-r-2}(e)^r \pi^{r-2}\,/a \\
&= 2^{r'+r+3} r^{-r-2} e^{r-4} \pi^{r-2}\, m
\end{aligned} \tag{5.11}$$

Thus there is an analogy between universes and particles. Universe energies exhibit a joint sequence in r and r'.

____

[21] c_g appears often in internal particle gambol models.

6. Fundamental Fermion Internal Universe Shells

In chapter 1 we considered the initial stage of the creation of universes. We saw that the mass-energy of a Parent universe emerged from the initial balance of internal mass-energy pressure and external Casimir force.

We now turn to consider a similar scenario for fundamental particles.[22] *We assume an elementary particle has an interior shell that is a miniscule universe.* We further assume it has a mass of a similar form to a universe given by an analogue of eq. 5.11. In the case of *our* universe, the universe external to the particle has dimension r' = 4 and the particle has an internal dimension r = 3. The form of the particle universe shell mass is

$$M_{r,r'} = 2^{\,r'+r+3}r^{-r-2}e^{r-4}\pi^{r-2}\,m$$
$$= 2^{\,r+7}r^{-r-2}e^{r-4}\pi^{r-2}m \tag{6.1}$$

where m is a mass parameter which we *initially* assume is common to all eight fundamental fermions.

---

[22] We considered similar possibilities in Blaha (2021d), (2022d) and (2023e).

7. Fermion Universe Shells and Internal Mass Sequences

This chapter is based on structuring fermions to consist of a universe with a mass defined by eq. 6.1. The particle universe consists of a "universe shell" contributing a mass factor to the total mass of a fermion, and of an interior mass factor. Their product equals the total mass of a fermion.

We start with the two sequences of fermions found earlier:

Sequence 2: v'   d   s   b
Sequence 1: e   u   c   t

The masses of the fermions in the sequences from Fig. 2.7 are:

<u>Sequence 2</u>

$$1 \qquad v' \simeq e2^{-15} \text{ GeV}/c^2$$

$$1 \qquad d = \pi e2^{-11} \text{ GeV}/c^2$$

$$\tfrac{3}{4} \qquad s = \tfrac{3}{4} \times \pi e2^{-6} \text{ GeV}/c^2$$

$$\tfrac{1}{2} \qquad b = \tfrac{1}{2} \times \pi e \text{ GeV}/c^2$$

<u>Sequence 1</u>

$$1 \qquad e = \pi e2^{-14} \text{ GeV}/c^2$$

$$\tfrac{3}{4} \qquad u = \tfrac{3}{4} \times \pi e2^{-12} \text{ GeV}/c^2$$

$$\tfrac{3}{4} \qquad c = \tfrac{3}{4} \times \pi^2 e2^{-4} \text{ GeV}/c^2$$

$$\tfrac{1}{2} \qquad t = \tfrac{1}{2} \times 4\pi^3 e \text{ GeV}/c^2$$

We interpret the above sequences' mass values as reflecting a product of an inner *universe shell* within each fermion and an internal fermion mass within the shell. We now consider this model for each fermion.

We view the sequence 1 fermions as having an internal universe shell of dimension r using eq. 6.1 modified to

$$M_{S,r,r'i} = 2^{r+7} r^{-r-2} e^{r-4} \pi^{r-2} m_{S0i} = c_{Shellr} \, m_{S0i} \qquad (7.1)$$

for sequence S = 1, 2 and for internal mass m_{S0i} with i = 1, 2, 3, 4 with

$$c_{Shellr} = 2^{r+7} r^{-r-2} e^{r-4} \pi^{r-2} \qquad (7.2)$$

Turning now to Sequence 2 fermions we find:

<u>Sequence 2 Fermion Masses</u>   (7.3)

$r = 3\ i = 1$ gives $M_{2,3,4v'} = v' \cong e2^{-15}$ GeV/$c^2 = 2^{10}3^{-5}\,e^{-1}\pi\ m_{20v'} = c_{Shell3}\ m_{20v'}$

$$m_{20v'} = v'/c_{Shell3}$$
$$= e2^{-15}/(2^{10}3^{-5}\,e^{-1}\pi)\ \text{GeV}/c^2$$
$$= (e^2/\pi)\,2^{-25}\,3^5\ \text{GeV}/c^2$$

$r = 3\ i = 2$ gives $M_{2,3,4d} = d = \pi e2^{-11}$ GeV/$c^2 = 2^{10}3^{-5}\,e^{-1}\pi\ m_{20d} = c_{Shell3}m_{20d}$

$$m_{20d} = d/c_{Shell3}$$
$$= \pi e2^{-11}/(2^{10}3^{-5}\,e^{-1}\pi)\ \text{GeV}/c^2$$
$$= e^2\,2^{-21}\,3^5\ \text{GeV}/c^2$$

$r = 3\ i = 3$ gives $M_{2,3,4s} = s = \tfrac{3}{4}\,\pi e2^{-6}$ GeV/$c^2 = 2^{10}3^{-5}\,e^{-1}\pi\ m_{20s} = c_{Shell3}\ m_{20s}$

$$m_{20s} = s/c_{Shell3}$$
$$= \tfrac{3}{4}\,\pi e\,2^{-6}/(2^{10}3^{-5}\,e^{-1}\pi)\ \text{GeV}/c^2$$
$$= \tfrac{3}{4}\,e^2\,2^{-16}\,3^5\ \text{GeV}/c^2$$

$r = 3\ i = 4$ gives $M_{2,3,4b} = b = \tfrac{1}{2}\,\pi e$ GeV/$c^2 = 2^{10}3^{-5}\,e^{-1}\pi\ m_{20b} = c_{Shell3}\ m_{20b}$

$$m_{20b} = b/c_{Shell3}$$
$$= \tfrac{1}{2}\,\pi e/(2^{10}3^{-5}\,e^{-1}\pi)\ \text{GeV}/c^2$$
$$= \tfrac{1}{2}\,e^2\,2^{-10}\,3^5\ \text{GeV}/c^2$$

These four sequence 2 fermions have an inner universe shell (a sphere) of 3 dimensions. Note fermion mass is denoted by its "letter." The four values have the approximate form:

$$m_{20i} = \alpha/\beta\ e^2 2^{-31+5i}\,3^5\ \text{GeV}/c^2 \tag{7.4}$$
$$= \alpha/\beta\ 2^{8+5i}\,3^5\,e^{-1}v_e$$

for $i = 1, 2, 3, 4$.

    We now consider sequence 1 fermions, which also has internal universe shells for values of r. We set $r = 3$ again.

<u>Sequence 1 Fermion Masses</u>   (7.5)

$r = 3,\ i = 4$ gives $M_{1,3,4t} = t = \tfrac{1}{2}\,4\pi^3 e$ GeV/$c^2 = 2^{10}3^{-5}\,e^{-1}\,\pi\ m_{10t} = c_{Shell3}\ m_{10t}$

$$m_{10t} = t/c_{Shell3}$$
$$= 2\pi^3 e/(2^{10}3^{-5}\,e^{-1}\pi) = e^2\pi^2\,2^{-9}\,3^5\ \text{GeV}/c^2$$

$r = 3,\ i = 3$ gives $M_{1,3,4c} = c = \tfrac{3}{4}\times\pi^2 e2^{-4}$ GeV/$c^2 = 2^{10}3^{-5}\,e^{-1}\pi\ m_{10c} = c_{Shell3}\ m_{10c}$

$$m_{10c} = c/c_{Shell3}$$
$$= \tfrac{3}{4}\,\pi^2 e\,2^{-4}/(2^{10}3^{-5}\,e^{-1}\pi) = \tfrac{3}{4}\,e^2\pi\,2^{-14}3^5\ \text{GeV}/c^2$$

$r = 3,\ i = 2$ gives $M_{1,3,4u} = u = \tfrac{3}{4}\times\pi e2^{-12}$ GeV/$c^2 = 2^{10}3^{-5}\,e^{-1}\pi\ m_{10u} = c_{Shell3}\ m_{10u}$

$$m_{10u} = u/c_{Shell3}$$
$$= \tfrac{3}{4}\,\pi e2^{-12}/(2^{10}3^{-5}\,e^{-1}\pi) = \tfrac{3}{4}\,e^2 2^{-22}\,3^5\ \text{GeV}/c^2$$

$r = 3.$ $i = 1$ gives $M_{1,3,4e} = \pi e 2^{-14}\, GeV/c^2 = 2^{10}3^{-5}\, e^{-1}\pi\, m_{10e} = c_{Shell3}\, m_{10e}$

$$m_{10e} = c/c_{Shell3}$$
$$= \pi e 2^{-14}/(2^{10}3^{-5}\, e^{-1}\pi) = e^2\, 2^{-24}3^5\, GeV/c^2$$

The four values have the approximate form:

$$m_{10i} = \alpha/\beta\ e^2\ \pi^{i-2}\, 2^{-29+5i}\ 3^5\ GeV/c^2 \tag{7.6}$$
$$= \alpha/\beta\ \pi^{i-2}\, 2^{10+5i}\ 3^5\ e^{-1}v_e$$

for $i = 1, 2, 3, 4$.

The ratio we calculate is

$$m_{10i}/m_{20i} = 4\, \pi^{i-2} \tag{7.7}$$
$$= M_{1,3,4i}/M_{2,3,4i}$$

Thus for b and t which both have $i = 4$:

$$M_{1,3,4t}/M_{2,3,4b} = 39.47 \tag{7.8}$$

while

$$t/b = 39.4$$

giving good agreement.

8. The Interior of a Particle Universe Shell

The masses of the particle interiors m_{10i} and m_{20i} of the sequences are given by eqs. 7.4 and 7.6. In each sequence they exhibit a dependence on an integer labeled i. We now treat those sequence numbers that specify each particle in each sequence and denote it as n:

$$m_{20n} = \alpha/\beta \; e^2 2^{-31 + 5n} \; 3^5 \; GeV/c^2 \tag{8.1}$$
$$= \alpha/\beta \; 2^{8 + 5n} \; 3^5 \; e^{-1} v_e$$
$$= \alpha/\beta \; (2^5)^n \; 2^8 \; 3^5 \; e^{-1} v_e$$

$$m_{10n} = \alpha/\beta \; e^2 \; \pi^{n-2} \; 2^{-29 + 5n} \; 3^5 \; GeV/c^2 \tag{8.2}$$
$$= \alpha/\beta \; \pi^{n-2} \; 2^{10 + 5n} \; 3^5 \; e^{-1} v_e$$
$$= \alpha/\beta \; (2^5\pi)^n \; 2^{10} \; 3^5 \pi^{-2} e^{-1} v_e$$

for n = 1, 2, 3, 4. We further introduce a "particle" entry into each sequence for the case of n = 0:

$$m_{200} = e^2 2^{-31} \; 3^5 \; GeV/c^2 \tag{8.3}$$
$$= 2^8 \; 3^5 \; e^{-1} v_e = 2.63 \times 10^{-6} \; GeV/c^2$$

$$m_{100} = e^2 \; \pi^{-2} \; 2^{-29} \; 3^5 \; GeV/c^2 \tag{8.4}$$
$$= 2^{10} \; 3^5 \; \pi^{-2} e^{-1} v_e = 5.08 \times 10^{-6} \; GeV/c^2$$

with $\alpha/\beta = 1$ by extending the particle list in Fig. 2.7 to the n = 0 case.

8.1 Derivation of the Form of Internal Fermion Masses

In section 1.6 we derived the form of the Coupling Constants. We found their form followed from the number of fermions in a fundamental representation of SU(n). A coupling constant $g(n) = g_n$ has the form:

$$g(n) = e \, (4\pi)^{\frac{1}{2}} \, 2^{n-6} = e \, (2^{-10}\pi)^{\frac{1}{2}} \, 2^n = 0.1505 \; 2^n \tag{1.18}$$
$$= g_g \; (\text{number of spin states per fermion})^{\text{number of fermions}}$$

We can now apply a similar logic to define the forms of m_{10n} and m_{20n}. In the case of sequence 2 we ascribe the factor of $(2^5)^n$ to the number of Normal fermions in one UST layer raised to the power of n:

$$(2^5)^n = (\text{number of Normal fermions in one layer})^n = 32^n \tag{8.5}$$

where n indicates the place of the fermion in sequence 2.

In the case of sequence 1 we approximate the factor of $(2^5\pi)^n \cong (2^7)^n$ where 2^7 is the total number of Normal fermions for all four UST layers. We raise this number to the power n:

$$(2^5\pi)^n \cong (2^7)^n = \text{(number of Normal fermions in all four layer)}^n = 128^n \qquad (8.6)$$

where n indicates the place of the fermion in sequence 1.

Thus we find that fermion masses build in multiples, $(2^5)^n$ or $(2^5\pi)^n$, of the previous fermion mass, fermion by fermion, in each sequence in a manner analogous to the growth of coupling constants, factor by factor of fermion spin **VALUE** $2 = 2\times\frac{1}{2} + 1$, for each coupling constant (eq. 1.18 above).

8.2 Comparison of Fermion Mass Sequences , Coupling Constant Sequence and Dimension Arrays

In this section we compare the fermion mass sequences with the coupling constant sequence and the sequence of Physical Cosmos dimension arrays. An interesting point is the equality of the number of entries in each sequence AND the four interaction coupling constants[23] that are relevant for the UST and particle physics as we know it. The below four coupling constants are all that is needed for the interactions of UST in all layers and generations. Because of the quadrupling effect as the Cosmos dimension r increases by 2, these four coupling constants are all that is needed for the four dimension array duplicates created each time r increases by 2. Similarly, the masses in other generations and layers may be expected to have analogous sequential patterns.

8.3 The Source of the Four-ness of these Features

In the 19th Century many sought for the source of the Nile River. Here we ask what is the source of the four-ness of the appearances of 4 in Cosmos Theory and the UST.

The answer is in the r = 0 Cosmos space where the dimension array has $2^4 = 4\times4 = 16$ elements. This dimension array is the seed for the quadrupling mechanism that we describe in our recent Cosmos books based on totally antisymmetric tensors and fractal sequences. The r = 0 dimension array is repeatedly quadrupled in higher dimensions r as one may see in Fig. 9.1 and other figures in our books.

Thus the mystery of four-ness is explained.

---

[23] We exclude the U(0) coupling constant since this group does not appear in the UST, and other higher dimension, Cosmos dimension arrays. See Fig. 9.1.

	Sequence 1					**Sequence 2**			
n:	1	2	3	4		1	2	3	4
Mass (GeV/c^2):	0.511×10^{-3}	1.80×10^{-3}	1.28	171		8.4×10^{-5}	4.24×10^{-3}	102×10^{-3}	4.34
Multiplier:		$2^5\pi$	$2^5\pi$	$2^5\pi$			$2^5=32$	$2^5=32$	$2^5=32$

		E S T I M A T E			**E X P E R I M E N T**			
n	**Interaction**	**Expression**	**g²/4π Value**	**g**	**Known[24] Value g²/4π**			**Deviation**
1	U(1) $\alpha_1 = \alpha$ Multiplier 4	$e^2/1024$	0.00721438	0.303	0.0072973525643			1.15%
2	SU(2) $\alpha_2 = g^2/4\pi$ Multiplier 4	$e^2/256$	0.0289	0.63	0.0316			9.3%
3	SU(3) $\alpha_3 = \alpha_S$ Multiplier 4	$e^2/64$	0.115	1.21	0.117			1.7%
4	SU(4)[25] α_4	$e^2/16$	0.462	2.4?	0.458			0.087%

Cosmos Dimension	0	2	4	6	8	10	12	14	16	18
Dimension Array Size[26]	4^2	8^2	16^2	32^2	64^2	128^2	256^2	512^2	1024^2	2048^2
Multiplier		4	4	4	4	4	4	4	4	4

Figure 8.1. The sequences of fermion masses (from Fig. 4.1) and coupling constants (from Fig. 1.3) and Cosmos Dimension Arrays. The four-ness is clearly seen.

8.4 Calculation of Universe Mass-Energy by Treating it as a Sequence 2 Particle

We can use eqs. 8.1 and 8.3 to determine a value for our universe's mass-energy $M_{universe}$ if we assume it has n = 64 – based on 128 Normal UST fermions – half of which are of sequence 2. We find a not unreasonable value:

$$M_{universe} = 32^{64} \times 2.63 \times 10^{-6} = 10^{90.75} \text{ GeV/c}^2 \qquad (8.7)$$

This estimated value of $M_{universe}$ is approximately $3.289\times10^9 \cong e^{-3}2^{36}$ times the mass of the observable (visible) mass part: $M_{observable} = 1.71 \times 10^{81}$ GeV/c^2.

---

[24] All coupling constant values are based on data from Particle Data Group Tables of 2024.

[25] This value is based on the "quadrupling trend" seen above in the three known coupling constants α_i.

[26] Dimension arrays are necessarily square just like γ-matrices since they both originate in the numerics of totally antisymmetric tensors. See Blaha (2024) *Fractal Cosmic Curve: Tensor-Based CosmosTheory* for details.

9. Vector and Scalar Boson Masses

This chapter deals with the Weinberg angle of ElectroWeak Theory and the massive vector boson and scalar boson (including Higgs bosons.) We show that the remarkable expressions and relations seen for Coupling Constants and fundamental fermion masses extends to these Unified Standard Theory (UST) features.

We show there are 256 scalar bosons. Some of these are Higgs bosons. Some are ordinary bosons. This difference raises the questions: Which bosons are of each type? What are their differentiating features? Appendix 9-A provides an answer to these questions based on the superluminal mass squared term for Higgs bosons and the subluminal mass squared term for non-Higgs bosons.

9.1 Weinberg Angle θ_W Calculation

The calculation of fermion mass data as functions of powers of 2 and π may be extended to the case of the vector boson masses in ElectroWeak theory. We previously showed the Weinberg angle θ_W, which is defined by

$$e^2/g^2 = \sin^2 \theta_W$$

where e is the electric charge. From Fig. 1.3 we find

$$\sin^2 \theta_W = \alpha_1/\alpha_2 = 256/1024 = 0.25$$
$$\sin \theta_W = 0.5$$

compared to the experimental value of $\sin^2 \theta_W = 0.231$. An accuracy to 8% that further supports the power sequence of Fig. 1.3.

9.2 Vector Boson Masses

Since the Weinberg angle is related to massive vector boson masses we see the conventional definition gives

$$m_W/m_Z = \cos \theta_W = 0.866 \tag{9.1}$$

in our approximation. They are known to have the values

$$W = 80.377 \text{ GeV}/c^2 \tag{9.2}$$
$$Z = 91.19 \text{ GeV}/c^2$$

where we use their symbols to represent their masses.

Noting that the t quark mass numerically satisfies

$$t/Z = 172.76 / 91.19 = 1.89 \cong 2 \qquad (9.3)$$

we see

$$Z = t/2 = 2^{15}\pi^3 \, v' \qquad (9.4)$$

thus giving an approximate expression for Z in terms of powers of 2 and π. Eq. 9.1 implies the W mass squared is

$$W^2 = 0.75Z^2 = \tfrac{3}{4} \, 2^{30}\pi^6 \, v'^2 \qquad (9.5)$$

or

$$W = \sqrt{\tfrac{3}{4}} \; 2^{15}\pi^3 \, v' \qquad (9.6)$$

Note the similarity to fermion mass relations:

$$Z/W = 1.13 \cong \pi/e = 1.16$$

Thus the W and Z masses have simple representations in terms of powers of 2 and π analogous to those of the fermion masses. There is a similar relation for the known Higgs particle mass.

9.3 UST/Cosmos Vector Boson Spectrum

The UST, which is for our universe with r = 4 dimensions, has a 256 dimension array for fundamental representations of symmetries. They are depicted in a four layers format in Fig. 9.1. The dimensions are allocated in the conventional way taking account of the nature of fundamental representation dimensions as complex dimensions. Thus eight real dimensions are allocated for the SU(4) Strong interaction group (later broken to SU(3)⊗U(1)); four real dimensions are allocated for the ElectroWeak group SU(2)⊗U(1); eight real dimensions are allocated for each Generation group; eight real dimensions are allocated for each Layer group; and four real dimensions are allocated for the space-time SO⁺(1,3). As a result the first layer top row occupies 16 dimensions for the Normal sector and 16 dimensions for the Dark sector. The first layer lower row occupies 16 dimensions for the Normal sector and 16 dimensions for the Dark sector.

The total number of dimensions allocated for the first layer symmetries is 64. Each of the other three layers has 64 dimensions allocated for their symmetry groups. The symmetry groups of each layer are different from those of other layers and thus independent of other layers. The total number of allocated dimensions is 256 as expected for the UST dimension array.

Note the ElectroWeak SU(2)⊗U(1) vector fields in the Normal sector of all four layers may be expected to be massive although the masses in each layer may well differ from those of other layers. A similar situation may prevail in the SU(2)⊗U(1) vector fields of the Dark sector.

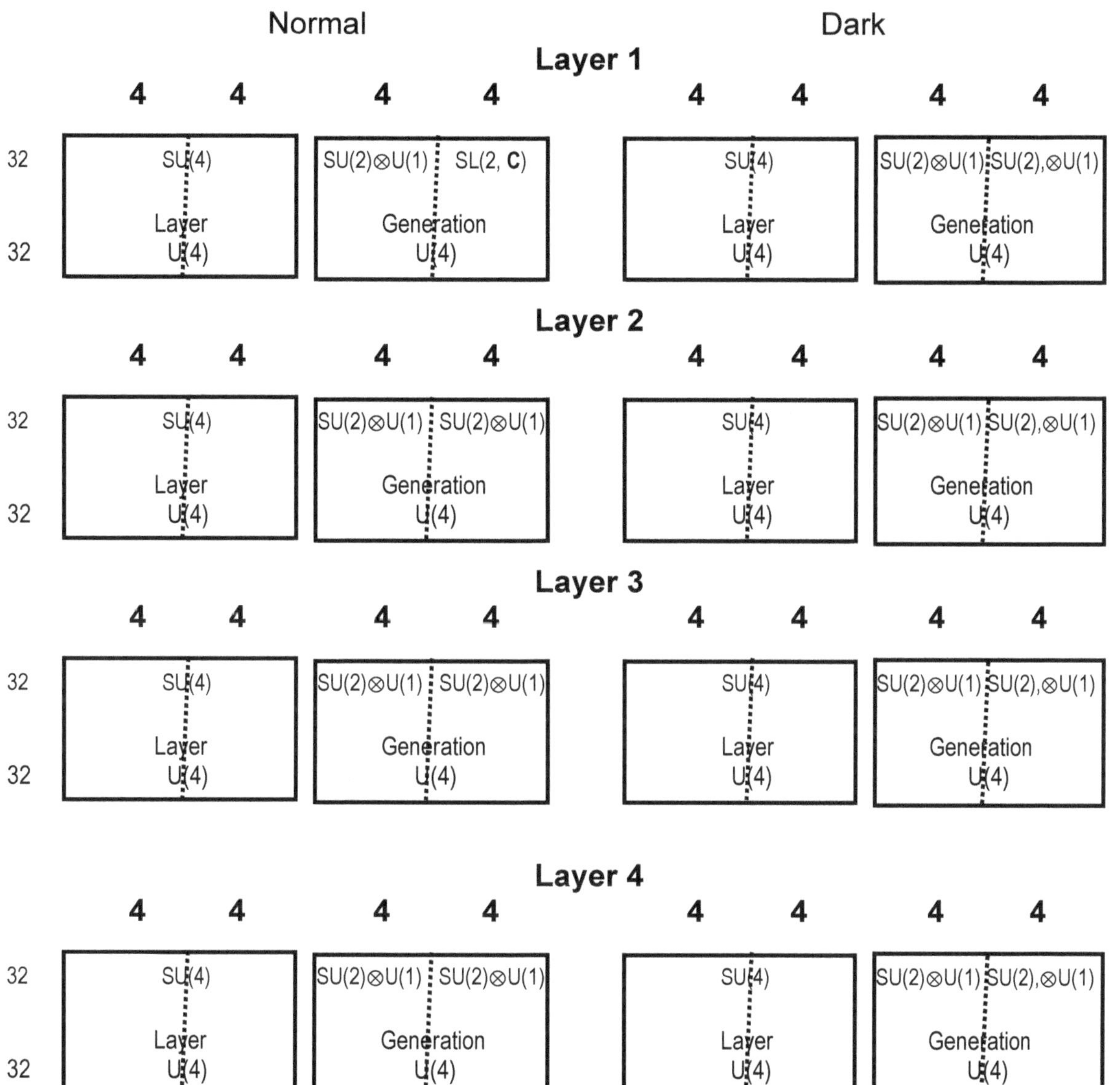

Figure 9.1 from Blaha (2020d). Normal and Dark symmetry groups of UST. SL(2, C) represents the Lorentz group $SO^+(1,3)$.

9.4 UST/Cosmos Scalar Boson Spectrum

The UST scalar bosons number 128 Normal bosons and 128 Dark bosons. They are subdivided into four layers with each consisting of four generations in a manner analogous to fermions. See section 9.5 below for a description of the scalar boson spectrum. In each scalar boson generation there will be Higgs bosons and non-Higgs bosons.

The separation of Higgs from non-Higgs bosons appears to be due to the su(1, 1) structure of the raising/lowering operators in Fourier expansions of boson quantum field operators. See Appendix 9-A below for details.

9.5 Scalar Boson Masses

We now determine the value of the known Higgs boson in terms of 2, e and π. The mass of the known SU(2) zero charge Higgs boson is

$$H = 125 \text{ GeV}/c^2 \qquad (9.7)$$

Since

$$H/Z = 1.37 \cong e/2 \qquad (9.8)$$

we may set

$$H = e2^{14}\pi^3 \, v' \qquad (9.9)$$

by eq. 9.4.

9.6 Scalar Boson Spectrum

The Higgs boson that we designate with H is a zero charge SU(2) scalar boson. There is also a unit charge boson in the SU(2) representation that we will denote as H_e. The mass of H_e is not known as yet although some experiments have tentatively seen more Higgs bosons.

There are 256 scalar bosons in the UST arranged in a manner similar to the 256 fundamental fermions in the UST. We will assume these bosons form a pattern (Fig. 9.2) similar to the fermion spectrum pattern in Fig. 2.2. We will assume **H is analogous to v'** and H_e is analogous to the electron.

We will further assume that the masses of these scalar bosons are related by a diagram (Fig. 9.3). Fig. 9.3 displays ratios denoted r_i where i =1, 2, ... , 7 that we will now define and use to generate relationships between masses. We define the ratios with their corresponding numeric values obtained directly from Fig. 9.3. See eqs. 2.1 – 2.8 for fermions. The scalar boson ratios are assumed to be the same based on the assumption that the internal structure of the scalar bosons is similar to the internal structure of fermions. In particular we assume the boson ratios are the same as the fermion ratios:

$$r_1 = e/\, v' = 6.08 \cong 2\pi \qquad (2.8)$$

corresponds to the H and H_e ratio r_{1b}:

$$r_{1b} = H_e \,/H = 6.08 \cong 2\pi \qquad (9.10)$$

Eq. 9.10 implies the unit charge partner of H has the very large mass[27]

$$H_e = 785 \text{ GeV/c}^2 \tag{9.11}$$

We set

$$H_e = e2^{15}\pi^4 v' \tag{9.12}$$

Other scalar boson masses are also expected to be very large.

We set the values of the eight layer 1, generation 1 boson ratios r_{ib} to those of the fermions in Fig. 2.8.

$r_{1b} = H_e/H = 6.08 \cong 2\pi$ $\qquad\qquad\qquad\qquad\qquad\qquad\qquad\qquad\qquad$ (9.13)
$r_{2b} \cong 4/\pi\ r_{1b} \cong 8$
$r_{3b} \cong r_{1b}/(4\pi) \cong 1/2$
$r_{4b} \cong 256\pi$
$r_{5b} \cong 1/(4\pi)$
$r_{6b} \cong 256\pi^2$
$r_{7b} \cong 1/(4\pi^2)$

Fig. 9.4 shows the resulting estimated approximate very large mass values.

---

[27] Andreas Crivellin *et al*, Nature Reviews Physics (2024) sees indications of a massive Higgs boson(s?) in the energy range from the proton mass through the teraelectronvolt scale. The 785 GeV/c^2 boson suggested above lies comfortably within that range.

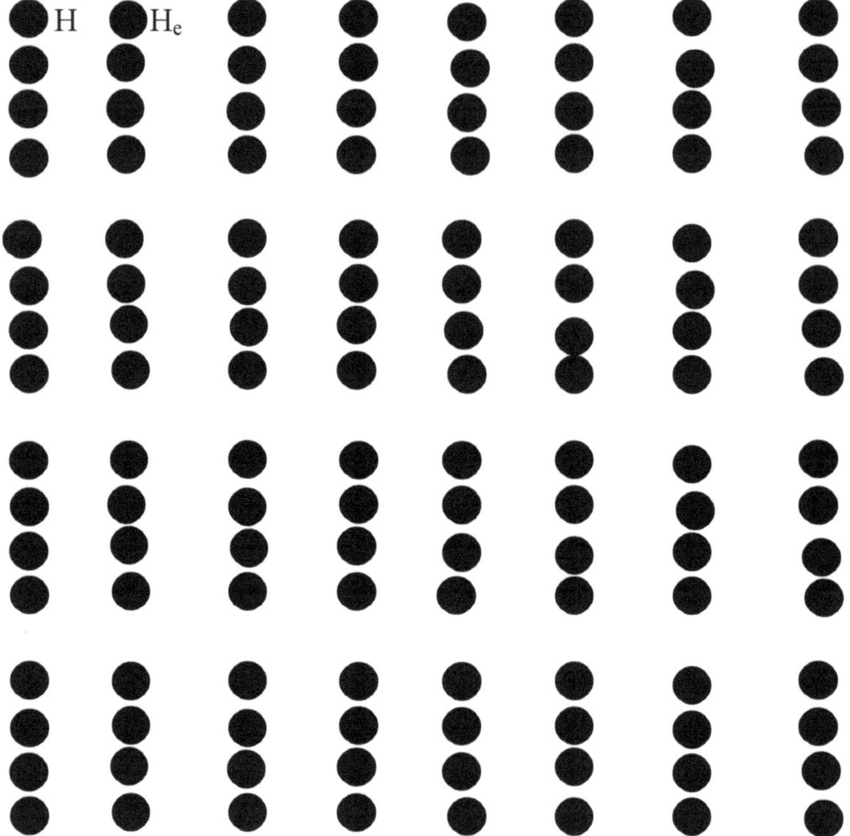

Figure 9.2. Four UST (and Cosmos) scalar boson layers with each consisting of four generations of eight normal and eight Dark bosons. Analogous to Fig. 2.2 for fermions.

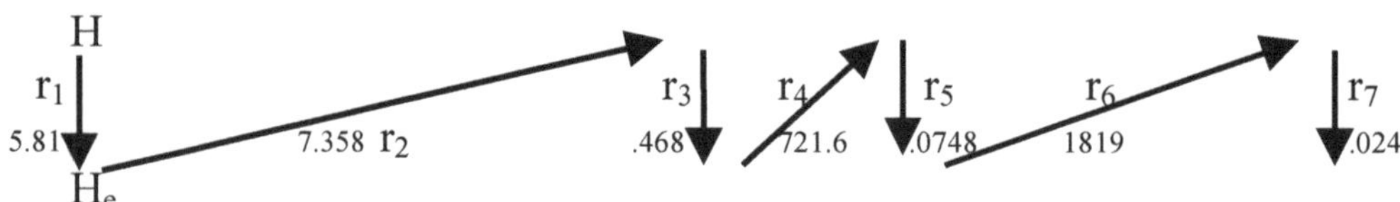

Figure 9.3. Fundamental first generation scalar boson masses. Note the charges and internal quantum numbers differ for bosons connected by arrows. We assume that the interior dynamics of the bosons are analogous to that of the fermions. Thus the mass ratios would be approximately the same.

ESTIMATE

$$H = e2^{14}\pi^3\, v' = 125$$
$$H_e = 2\pi\, H = 785$$
$$H_d = 2^4\pi\, H = 6283$$
$$H_u = 2^3\pi\, H = 3142$$
$$H_c = 2^{11}\pi^2 H = 2.5 \times 10^6$$
$$H_s = 2^9\pi\, H = 201{,}062$$
$$H_t = 2^{17}\pi^3 H = 5.08 \times 10^8$$
$$H_b = 2^{15}\pi H = 12.87 \times 10^6$$

Figure 9.4. Approximate estimates of the eight layer 1, generation 1 boson masses in GeV/c^2. The subscript a on the ratios H_a indicates the corresponding fermion.

Appendix 9-A. Separation of Scalar Bosons into Higgs and Non-Higgs Sectors

This appendix is chapter 4 from Blaha (2023d). It suggests a reason for the separation of scalar bosons into Higgs bosons and non-Higgs bosons based on the superluminal nature of su(1, 1) Higgs bosons. The superluminal – subluminal distinction is in su(1, 1) representations of raising and lowering operators of Fourier expanded boson wave functions.

4. FRF Contents Determined by Two Times in HyperUnification Space

The existence of Superluminal dynamics is an existing issue that has been theoretically studied over a period of years. We studied superluminal Quantum Field Theory a number of times. Blaha (2007a) describes superluminal Quantum Field in some detail and shows it defines a completely reasonable dynamics. Blaha (2018e) studies superluminal Statistical Mechanics and Thermodynamics and shows that they conform to Physical expectations such as Maxwell-Boltzmann theory and the laws of Thermodynamics.

In this chapter we consider the role of superluminal/sublight considerations in the determination of the structure of fermions and Symmetry Groups of *Fundamental Reference Frames* (FRF's) and in the form of Higgs and other scalar particles.

4.1 The Higgs Mechanism – Quark Confinement Dichotomy

A possibly deeper view of the Higgs Mechanism is based on a Dichotomy between the Higgs Mechanism and Quark confinement. *The Higgs Mechanism and the confining Strong Interaction are alternatives.* The Higgs Mechanism gives masses to particles. Quark Confinement is based on an effective infinite potential (mass) energy that prevents quark deconfinement.

The Higgs Mechanism uses a tachyonic scalar particle implementation[28] for symmetry breaking because it works despite its being superluminal due to negative mass squared. The Strong Interaction has a non-Higgs, non-tachyonic confinement mechanism.[29] These mechanisms appear interrelated.

A simple boson example illustrating the dichotomy in the mechanisms starts with the PseudoQuantum scalar field equations:

---

[28] The tachyonic nature is removed by quartic interactions.
[29] See Blaha (2022f) for a detailed discussion of quark confinement and of related MOND-like gravitation.

$$\Box\varphi_1 - m^2\varphi_1 + g\varphi_1{}^4 = 0 \qquad\qquad (4.1)$$
$$\Box\varphi_2 - m^2\varphi_2 + g\varphi_2{}^4 = 0 \qquad\qquad (4.2)$$

with tachyonic negative mass squared. Both the type 1 and type 2 fields have the same field equation as we have seen in earlier work.

On the other hand, the prototype Strong Interaction PseudoQuantum scalar field equations:

$$\Box\varphi_1 \pm m^2\varphi_2 = 0 \qquad\qquad (4.3)$$
$$\Box\varphi_2 = J \qquad\qquad (4.4)$$

with the quartic interaction dropped for simplicity yield

$$\Box^2\varphi_1 \pm m^2 J = 0 \qquad\qquad (4.5)$$

The differential operator $\Box^2$ is quartic. Choosing the minus sign to have a positive J source term: results in

$$\Box^2\varphi_1 = m^2 J \qquad\qquad (4.6)$$

which specifies a confining linear r potential when used in a gauge field theory.[30]

Note the minus sign places eqs. 4.1 and 4.3 on the same footing with the difference that eq. 4.1 is tachyonic but eq.4.3 is not tachyonic due to the appearance of the type 2 field in the eq. 4.3 mass term. Tachyon behavior is avoided in order to obtain Strong interaction confinement.

We called the difference between the Strong Interaction mechanism and the Higgs Mechanism the *Higgs – Confinement Dichotomy* in our earlier books.

4.2 Sublight vs. Superluminal Sectors in su(1,1) and F-Theory

F-Theory and su(1,1), which appear in the HyperCosmos and the Second Kind HyperCosmos, have two time dimensions that lead to a more intricate form of sublight-superluminal separation. These separations lead to the split of symmetries seen in The Standard Model, the UST, and in HyperCosmos spaces. The su(1, 1) group appears as a subspace in all spaces $N \leq 7$ above our $N = 7$ space.

This phenomenon together with the structuring imposed by Cosmos Theory almost completely specifies the separations of Internal Symmetry in the Standard Model and its extension in the UST and the HyperCosmos spaces.

The split is evident in the contents of the symmetry group FRF content of each Cosmos Theory space. Together with the spaces, HyperUnification spaces and FRFs development the total structure of the Cosmos Theory universes (including our own UST universe) is determined.

We now proceed to determine the separation of FRF content from superluminal/subluminal considerations.

---

[30] See Stephen Blaha, "New Framework for Gauge Field Theories", IL Nuovo Cimento **49A**, 113 (1979) for an SU(3) gauge theory version with confinement.

4.3 Two Times Coordinates and Fermion Separations

An su(1, 1) group coordinate space has a metric with two real time coordinates:

$$ds^2 = t_{01}{}^2 + t_{02}{}^2 - x_1{}^2 - x_2{}^2 \qquad (4.7)$$

In Blaha (2007b) we showed that the four species of fermions: e-type, v-type, q-up-type and q-down-type followed from the four types of boosts of the complexified Lorentz Group $SO^+(1,3)$, which is often expressed as a representation of SL(2, **C**)). Complex Lorentz group boosts separate into sublight, superluminal, complex sublight, and complex superluminal boosts yielding the four fermion species respectively.

Now we have a more interesting situation with two time coordinates. Each has its own light speed. We take them to be numerically equal, although different "coordinate system-wise", for the purposes of our discussion. One light speed divides the set of possible fermions into two "superspecies", namely Normal and Dark fermions. The second light speed further divides each superspecies into four parts. The result is the eight fermion species in our universe (counting each quark as a different species). It includes both Normal and Dark sectors. It is also specifies the form of fermion species in the other HyperCosmos, and Second Kind HyperCosmos, spaces, all of which also have su(1, 1) multiple time coordinates.

The 16 fermions of the N = 7 Fundamental Reference Frame (our universe's) can be separated into a pair of eight subspecies if account is taken of the occurrence of each quark species as a triplet. This separation of fermion species leads to the structure of the Unified SuperStandard Theory (UST) and the Standard Model. It also leads to the form of Fundamental Reference Frames for all HyperCosmos (and 2nd Kind HyperCosmos) spaces.

The FRFs of higher r space-time dimensions have replicates of these 16 fundamental fermions. All fermions in the replicates are different with differing quantum numbers.

In the case of 2nd Kind HyperCosmos spaces we find their FRF's have multiples of only the sublight t_{01} Normal fermions. All the fermions in their higher space-time FRF replicates are different with differing quantum numbers. The Dark sector is missing in all 2nd HyperCosmos FRF's and spaces.

A complete, detailed spectrum of fundamental fermions in our universe, and other universes of all Cosmos Theory spaces, thus emerges, part of which appears in the UST and Standard Model for our universe.

t_{01}:	sublight **Normal**		superluminal **Dark**	
t_{02}:	sublight	superluminal	sublight	superluminal
	q-up-type and q-down-type	e-type and v-type	q-up-type and q-down-type	e-type and v-type
Number of fermions:	6	2	6	2

Figure 4.1. Separation of fundamental fermions in the N = 7 space's 16 dimension FRF. The superluminality of fermions is symbolic since the su(1, 1) coordinates are not space-time coordinates of our universe.

4.4 Scalar Bosons Corresponding to FRF Dimensions

We now consider the separation of scalar bosons including Higgs bosons when mapped from an FRF's dimensions. This case is particularly interesting because it evidences the separation effects of the two time coordinates which lead to sublight and superluminal particle differences.

Unlike the fermion case the irreducible representations of the symmetry groups have complex dimensions that may be separated into real and imaginary parts. The N = 7 space FRF maps to a set of 16 scalar bosons. These bosons are first separated by time coordinate t_{01} into Normal and Dark bosons. Then t_{02} separates the bosons in each sector into Higgs and non-Higgs bosons mirroring the fermion case above. See Fig. 4.2.

In the N = 7 space FRF Normal sector there is a non-Higgs set of three SU(3) bosons corresponding to the up-type and down-type quarks – totaling 6 bosons , and two Higgs bosons forming an SU(2) representation. The Dark sector has similar sets of bosons.

t_{01}:	sublight **Normal**		superluminal **Dark**	
Bosons:	8		8	
t_{02}:	sublight	superluminal	sublight	superluminal
	Non-Higgs	Higgs Bosons	Non-Higgs	Higgs Bosons
Bosons:	6	2	6	2
Boson Group	SU(3)	SU(2)	SU(3)	SU(2)

Figure 4.2. Bosons separated by su(1, 1)'s two times coordinates – not space-time coordinates. There is one boson per fermion. Thus 6 bosons corresponding to the 6 quarks and two bosons corresponding to e and v. Their group memberships are displayed as well. Bosons can be mapped from the FRF as well.

The N = 7 space FRF is mapped to a set of replicates in each FRF of higher space-time dimensions. Thus the set of split Fig. 4.2 bosons map to scalar bosons replicate sets. All bosons have separate quantum numbers and lay in different symmetry group representations.

4.5 Internal Symmetries of the HyperCosmos for N = 7

This section shows the N = 7 space FRF separation of internal symmetries due to the two time coordinates of su(1,1) – a group that appears in all HyperCosmos and all Second Kind HyperCosmos spaces.

t_{01}:	sublight **Normal**		superluminal **Dark**	
Real-valued Dimensions	U(4)		U(4)	
	8		8	
t_{02}:	sublight	superluminal	sublight	superluminal
	6	2	6	2
	SU(3) Confimement	U(1)	SU(3) Confimement	U(1)
OR		Higgs Mechanism		Higgs Mechanism
	4	4	4	4
	$SO^+(1,3)$	SU(2)⊗U(1)	SU(2)	SU(2)⊗U(1)
OR				
	4	4	4	4
	SU(2)	SU(2)⊗U(1)	SU(2)	SU(2)⊗U(1)

Figure 4.3. Split FRF dimensions mapped as real-valued representation dimensions to one of U(4)⊗U(4) for Generation and Layer symmetry groups, or $[SU(3)⊗U(1)]^2$ or $SO^+(1,3)⊗SU(2)⊗[SU(2)⊗U(1)]^2$ or $[SU(2)⊗SU(2)⊗U(1)]^2$ The SU(2) groups are Connection Groups. The SU(2)⊗U(1) groups are ElectroWeak Groups. See Fig. 2.6 for the N = 7 case.

4.6 Internal Symmetries of the HyperCosmos for N Greater than 7

The FRF's for higher space-time dimensions consist of replicates of the N = 7 dimensions. Thus the Symmetry Groups of these FRF's are replicates of the symmetry group maps in Fig. 4.3. The r = 6 FRF has two replicates (see eq. 3.8) of the N = 7 FRF. The symmetry groups of all replicates are different with different sets of quantum numbers. The replicates then map through GR transformations to the set of replicates embodied in the space's dimension array.

Comments

Note the sublight SU(3) group has a confining Strong interaction generated by a higher derivative interaction. Gravitation which should also be treated as sublight also has a higher derivative interaction that produces a MOND-like gravitational potential. See chapters 12 – 14 of Blaha (2022f) for details.

The split of particles and groups is based on the superluminal/subluminal separation of the coordinates of eq. 4.7 of the su(1,1) group embedded within the creation/annihilation operators of fermion wave functions. It is not based on universe space-time coordinates.

The values of the nonzero dimensions of an FRF have no Physical meaning. The dimensions acquire Physical meaning only when mapped to particles and symmetry groups as above.

The values of the dimensions in the dimension array of a Cosmos space also have no Physical meaning. They acquire Physical meaning when mapped to particles and symmetry groups in universe creation.

The value of the dimension in a one[31] dimension FRF has no Physical meaning. Physical meaning is acquired after it is transformed to a space's dimension array and mapped to particles and symmetry groups in universe creation.

The Coupling Constants, the dimension arrays, and the relation of coupling constants to masses (and thereby to energies) all reflect their common interpretation as manifestations of degrees of freedom. There is a unifying principle at the basis of Cosmos Theory in its spaces, symmetries, universes, universe structure, energies, and particle interactions.

---

[31] One might call the dimension the "Adam" dimension since it is the source of all dimension array dimensions.

10. Generations of Fermions

We have suggested earlier that the Cosmos Theory portrayal of generations of fermions should be preferred especially in view of the patterns of first generation current fermion masses.

We now consider the extension of the first generation mass patterns and spectrum to the second and third generations.[32] The masses of the μ and τ leptons (of the second and third generations) are known. We can use their mass ratios to *approximately* extend the quark and neutrino masses to the second and third generations.

The μ and τ mass ratios can be approximated as

$$\mu/e = 207 \cong 2\pi^4 = 195$$
$$\tau/\mu = 16.6 \cong 2^4 = 16$$

where

$$\mu = 0.106 \text{ GeV/c}^2$$
$$\tau = 1.777 \text{ GeV/c}^2$$

Based on these ratios we take the estimates in Figs. 2.4 and 4.1 and determine the second and third generation fermion masses using the fermion sequences of Fig. 4.1. For the second generation we use the ν'/e ratio to find

$$\nu_2' = \nu'\mu/e = 1.7\times10^{-2} \text{ GeV/c}^2$$

Then we use the multipliers for each sequence to obtain the masses of the other fermions in the sequence for generations 2 and 3.

The process determines the second and third neutrino masses from the ratios of the charged leptons. Then each sequence of masses is generated from the sequence's multiplier starting with the sequence's lepton mass.

---

[32] Fourth generation masses remain to be determined when experimental data becomes available.

First Generation

| | SEQUENCE 1 | | | | SEQUENCE 2 | | |

	e	u	c	t	ν'	d	s	b
Mass:	0.511×10^{-3}	1.80×10^{-3}	1.28	171	8.4×10^{-5}	4.24×10^{-3}	102×10^{-3}	4.34
Multiplier:		$2^5\pi$	$2^5\pi$	$2^5\pi$		32	32	32

Sequence 1 multiplier: $2^5\pi = 100.5 \cong 100$; Sequence 2 multiplier 32. Masses in GeV/c^2

Figure 10.1. First Generation fermion masses. Masses in GeV/c^2.

Second Generation

Ratio: $207 \cong 2\pi^4 = 195$ for μ and ν_2'

	μ	u_2	c_2	t_2	ν_2'	d_2	s_2	b_2
	0.106	10.6	1060	10^5	1.64×10^{-2}	52×10^{-2}	1677	53673
	$2^5\pi$	$2^5\pi$	$2^5\pi$		32	32	32	

Figure 10.2. Projected approximate Second Generation fermion masses. Masses in GeV/c^2.

We now set $\nu_3' = \nu_2'\, \tau/\mu = 0.272\ GeV/c^2$.

Third Generation

Ratio: $\tau/\mu = 16.6 \cong 2^4 = 16$ for τ and ν_3'

| | SEQUENCE 1 | | | | SEQUENCE 2 | | |

	τ	u_3	c_3	t_3	ν_3'	d_3	s_3	b_3
	1.777	178	17800	1.78×10^6	0.262	8.4	269	8598
	100	100	100		32	32	32	

Note: The physical neutrino mass is $\nu_3 = \nu_3'\, U(0)^2/e^2 = \nu_3' e^2/4096^2 = 1.2\times10^{-7}\ GeV/c^2$ = 120 ev/c^2

Figure 10.3. Projected approximate Third Generation fermion masses. Masses in GeV/c^2.

11. Higgs Mechanism and Cosmos Particle Masses

The Cosmos Theory sequences of fermion masses was based on an analysis of fundamental fermion masses. Higgs Models are also based on fits to experimental mass data. Therefore they cannot conflict because of their common origin in the fermion mass spectrum.

The accuracy of the Cosmos analysis masses, and the assumed accuracy of Higgs Model fits for masses, imply a relation between Higgs Model parameters and the Cosmos fits. Since they both have the same phenomenological source *there can be no conflict* in their respective results.

It is therefore a matter of taste as to which to select since both Cosmos masses and Higgs masses are phenomenologically equivalent.[33]

Cosmos Theory and its coupling constants and masses reflect an underlying theoretical framework that embodies the Reality that might be present in a "true" dynamic quantum field theory of space, matter, and energy.

The fundamental theoretical framework of Reality must be independent of space and time, and thus is not a familiar type of dynamical theory. It seems to be combinatoric in form since it is based on integer relations, and a few fundamental constants e, and π.

Appendix 9-A provides a possible answer to the questions: What vector bosons acquire mass through the Higgs Mechanism? What distinguishes Higgs bosons from non-Higgs bosons? (The UST and Cosmos Theory indicate there are 256 scalar bosons in our universe.)

The origin of mass seems to be a result of the neutrino's mass. Ultimately all fermion masses (and thereby boson masses) seem to follow from the sequences of fermion masses of earlier chapters 2, 4, 6, 7 and 8. All masses are derivable from the electron neutrino mass. Particles may be viewed as miniscule universes. Universes follow from consistency conditions that naturally embody a mass parameter on simple dimensional grounds. Thus Mass has an ultimate origin in the consistency conditions for a Parent universe and for a particle universe. The masses of particles and universes may be viewed as "slices" of mass.[34] They are symbolically represented on the book's cover.

---

[33] A map between the Cosmos patterns of masses and Higgs Model parameters would be of interest.
[34] See chapter 12. Universes can also be viewed as composed of slices. See *Cosmos Theory: The Sub-Particle Gambol Model* and *Cosmos-Universe-Particle-Gambol Theory*.

12. Gambol Derivation from Fundamental Fermion Internals

The multiplicative factors in the two fermion sequences imply that each fermion mass is a multiple of the previous fermion mass in the sequence:

$$m_n = \pi 2^5 m_{n-1} \quad \text{Sequence 1} \qquad (12.1)$$
$$m_n = 2^5 m_{n-1} \quad \text{Sequence 2} \qquad (12.2)$$

Eqs. 12.1 and 12.2 show that the masses of the fundamental fermions in each sequence grow by a *multiplicative* factor of $\pi 2^5$ (sequence 1) or 2^5 (sequence 2) fermion by fermion. Each fermion has $\pi 2^5$ (sequence 1) or 2^5 (sequence 2) factors of the preceding fermion in the sequence.

This interpretation is a realization of the gambol concept introduced by the author[35] previously. A fermion or hadron is viewed as composed of s parts where s is an integer. Its mass m_g is 1/s of the particle mass m:

$$m_g = m/s \qquad (12.3)$$

Comparing eqs. 12.1, 12.2 and 12.3 we see we can view

$$m_g = m/(\pi 2^5) = m_{n-1} \text{ with } m = m_n \qquad \text{Sequence 1} \quad (12.4)$$
$$m_g = m/2^5 = m_{n-1} \text{ with } m = m_n \qquad \text{Sequence 2}$$

Then the gambol mass is 2^{-5} times the mass of the preceding fermion for sequence 2 and $2^{-5}/\pi$ times the mass of the preceding fermion for sequence 1.

12.1 Repeated Fractionation of Fermions

A sequence 2 fermion may be separated into 32 slices as we see symbolically represented in Figs. 12.1 and 12.2. Each slice may in turn be separated into 32 slices of the next lower fermion and so on. The separation into slices is bounded below by the length of the fermion sequence.

We can extend the sequence of mass slicing mathematically indefinitely. Thus one obtains gambol modeling through fractionating by powers of $s = 32$.

Similar comments apply to sequence 1 fermions with the slicing in $\pi 2^5$ parts.

---

[35] See Blaha (2023e) *Cosmos Theory: The Sub-Particle Gambol Model* and (2024a) *Cosmos-Universe-Particle-Gambol Theory.*

12.2 Comparison of Slicing to Coupling Constant Slicing

The slicing of fermion masses shown in Figs. 12.1 and 12.2 is very similar to the factorization of coupling constants seen in section 1.6. There may be a connection between them in a deeper analysis of fermion dynamics.

12.3 Combined Fermion Slices

Sequence 2 slicing may be viewed as generated by fractionation. They may be combined in sets of 16, 8, 4, and 2 slices. *Each combined slice not only has a mass that is the sum of the individual slice masses but it also has the spin and internal symmetry quantum numbers of the fermion (or hadron) of which it is a part.* We show this feature in the case of deep inelastic lepton-nucleon scattering in Appendix 12-A.

The gambol models of fundamental fermions, and their behavior, can be extended to hadrons, such as nucleons and bound states such as we have done in earlier work. The gambols (slices) have the masses set by fractionation and their behavior under interactions set by the spin and internal symmetries of the particle of which they are a part. See section 12.4 for references.

Experimental data is becoming available showing that particle spin and internal symmetries can be separated from the inherent features of the particle. This data accords well with Gambol Theory.

12.4 Proof of Gambol Modeling of Sequence 2 Fermions

The effect of section 12.3 is to prove the "gamboling" of sequence 2 fermions in fractions of $s = 2, 4, 8, \ldots, \infty$ as we saw in earlier work by this author. Each gambol may be further subdivided into 32 parts and so on following the order of sequence 2 fermions. See Fig. 12.1 for a symbolic representation of a single factoring (gamboling) into 32 parts. See the book's cover for a *symbolic*[36] depiction of a sequence of nested gambols.

The choice of the value of the gambol s value for a fermion can be generalized to all even integers resulting in Gambol Theory. It can be further generalized to all real numbers (especially in view of the non-integer multiplier for sequence 1.)

In our past studies of gambol models of particles we typically set $s = 8$. We now see that $s = 8$ implies four combined 1/32 slices. Thus our earlier work stands as presented.

We can now consider factoring a quark mass into gambol masses. For example we can factor the d quark into 32 parts with

$$m_{gd} = m_d/(32) \tag{12.5}$$

and proceed to consider applications in the Gambol Theory of quarks.

Sequence 2 fermions support a gambol view of particles. The gambol view is based on the extension of s to all even integers. Thus we have found the door to the interior of particles from the consideration of sequence 2 fermion masses found

---

[36] The cover depiction *should* have shown 32 part subdivisions of each of the 32 parts.

experimentally. This approach is distinguished from the approach of Weyl, Einstein, and others to the internals of elementary particles based on theoretical considerations.

12.5 Proof of Gambol Modeling of Sequence 1 Fermions

The proof is similar to that of sequence 2 fermions. However the gambol s value must be taken to be any real valued number since the sequence 1 multiplier is $\pi 2^5$. We can set the value of $s = \pi 2^5$ initially. Then one can set $s = (\pi 2^5)^n$ for sequence 1 fermions. Then we generalize s to any real value. A sequence 1 fermion may be factored into s_0 parts:

$$m_g = m/s_0 \tag{12.6}$$

Fig. 12.2 is a symbolic example of sequence 1 fermion factoring into gambols.

12.6 Proof of Gambol Modeling of Hadrons

The above proofs of the basis of Gambol Theory for fundamental fermions can be directly generalized to hadrons. A hadron of mass M may be gamboled into parts of mass m_g. In interactions each gambol has the spin and internal symmetry quantum numbers of the parent hadron.

$$m_g = M/s_0 \tag{12.7}$$

We used this approach in deep inelastic lepton – nucleon scattering. See Appendix 12-A. Section 12.7 lists our other successful applications.

Figure 12.1. Sequence 2 fractionation of a particle into 32 gambols. Each gambol is a copy of $(2^5)^{-1}$ times the mass and structure (but not the spin or internal quantum numbers of the particle. The gambol acquires these aspects by inheritance from the particle. Each gambol acts like the particle except for gambol mass and gambol temperature.

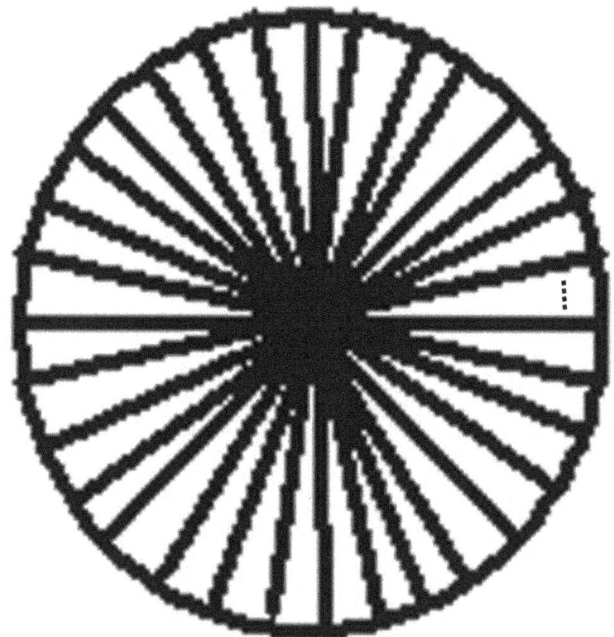

Figure 12.2. Sequence 1 fractionation of a particle into 32π gambols. Each gambol is a copy of $(\pi 2^5)^{-1}$ times the mass and structure (but not the spin or internal quantum numbers) of the particle. The gambol acquires these aspects by inheritance from the particle. Each gambol acts like the particle except for gambol mass and gambol temperature.

12.7 Gambol Models of Fermions and Hadrons

Much of our gambol modeling appears in the books:

Cosmos Theory: The Sub-Particle Gambol Model
Describes its use in deep inelastic lepton and quark resonances, production and interactions, hadron-hadron collisions; neutrino oscillations e - μ, μ – τ; particle stability pressure vs. Casimir force, and Hubble expansion. See section 9.8 and 9.9. In section 9.9 (eq. 13.22) we find an approximate relation, which we now express as

$$c_g = 0.0785 \qquad (12.8)$$

giving a 2% difference from the approximate relation we found $c_g = 0.0785$.

Cosmos-Universe-Particle-Gambol Theory
Here we found the following uses of c_g:

The Cosmos Theory Distribution of Universes
Anti-Planckian Model of Hubble Expansion Parameter
Our Universe as a Megaverse Particle
Universe-Megaverse Boundary
Casimir Force Limit on Hubble Expansion
S_8 Universe-Gambol Clumping Models
Variation of Universe Cluster Mass-Energy Sizes
Neutron Star Quark Core Gambol Model
GVDM Photon – ρ meson Gambol Model
Higgs and Massive Vector Boson Gambol Models

12.8 A New Constant of Nature c_g

The gambol models of fundamental fermions and hadrons define a gambol temperature T with the relation:

$$kT = c_g M = 0.0785M \qquad (12.9)$$

where M is the mass of the particle. This temperature appears in our gambol Planckian models of fermions and hadrons.

The author has introduced a new physical constant

$$c_g = 0.0785 \cong \pi^2/128 = 0.0771 \qquad (12.10)$$

(accurate approximately to 1.8%) that appears in many contexts in our studies in earlier books such as Blaha (2023e) and (2024a).

12.9 Summary of Proof of Gambol Theory

The proof of Gambol Theory begins with the experimentally based fundamental fermion sequences. They show that the mass of each fermion is a multiple of nested fermion layer masses starting from the next lower fermion in the sequence.

We generalize this factoring of each fermion to arbitrary factorings using the gambol factoring value s which may be an integer or a real valued number. The subdivided fermion has gambols within it.

Gambol Theory generalizes the factoring process to give the slices of a fermion the spin and internal symmetries of the parent fermion. This feature is exhibited in

Appendix 12-A for deep inelastic lepton – nucleon scattering. It is also exhibited in the cases presented in the books listed in section 12.7.

We further generalize Gambol Theory to hadrons, which may be partitioned into gambols. Some of the cases in section 12.7 reflect hadron gamboling. There are also applications of Gambol Theory to universes.

We find Gambol Theory may well have universal significance in the dynamics of particles and universes.

Appendix 12-A. Application of Gambol Dynamics to Deep Inelastic Electron-Proton Scattering

This appendix is chapter 6 of Blaha (2023a).

The picture of gambol dynamics that we have developed has physical applications to particle interactions that we now explore. In this chapter we will use the parton model[37] of deep inelastic electron-proton scattering[38] as a starting point to develop a detailed model of gambols that appears to have the possibility of wide application in analyzing particle interactions. We will see that deep inelastic scattering data suggest that fundamental particles have a probabilistic spectrum of gambols within them that have a Planck-like black body distribution. In subsequent chapters[39] we will apply this model to particle decay – specifically heavy quark, muon, tau, neutrino and pion decays.

6.1 Parton Model of Deep Inelastic Lepton-Nucleon Scattering

Fig. 6.1 displays the parton deep inelastic scattering process. An electron interacts with a nucleon (composed of partons) of momentum P in an infinite momentum reference frame via the exchange of a virtual photon of momentum q. The nucleon is viewed as a composed of partons. The variables are;

$$Q^2 = -q^2 \tag{6.1}$$
$$\nu = q{\cdot}P/M$$

where M is the nucleon mass. The experimental data shown in Fig. 6.2 exhibits approximate scaling in a variable denoted

$$x = Q^2/(2M\nu) \tag{6.2}$$

Fig. 6.2 shows a plot of the deep inelastic e-p structure function $F(\omega) = \nu W_2(\nu, q^2)$ vs. the parameter ω that shows scaling in $x = 1/(2M\omega)$ where

$$\omega = 1/(2Mx) \tag{6.3}$$

Bjorken and Paschos define an expression for W_2:

---

[37] J. D. Bjorken and E. A. Paschos, Phys. Rev. 185, 1975 (1969).
[38] Stephen Blaha, Phys. Rev. D**3**, 510 (1971) and Phys. Lett. B **40**, 501 (1972).
[39] See Blaha (2022d) for an initial description of these gambol applications.

$$W_2(\nu, q^2) = \Sigma_N \, P(N) \, (\Sigma_i \, Q_i^2)_N \int_0^1 dx \, f_N(x) \, \delta(\nu - Q^2/(2xM)) \qquad (6.4)$$

which, after integration, gives[40]

$$F(\omega) = \nu W_2(\nu, q^2) = \Sigma_N \, P(N) \, (\Sigma_i \, Q_i^2)_N \, f_N(x) \qquad (6.5)$$

with x given by eq. 6.2 and ω given by eq. 6.3. The function P(N) is the probability of an N parton state in the proton. The $(\Sigma_i \, Q_i^2)_N$ is the average of the charges squared in the N parton state. The function $f_N(x)$ is the model dependent probability of finding a parton of longitudinal momentum fraction xP in the N parton state.

Eq. 6.5 corresponds to the experimental data in Fig. 6.2. Fig. 6.2 shows a plot of the deep inelastic e-p structure function $F(\omega) = \nu W_2(\nu, q^2)$ vs. the parameter ω showing scaling in x = 1/(2Mω).

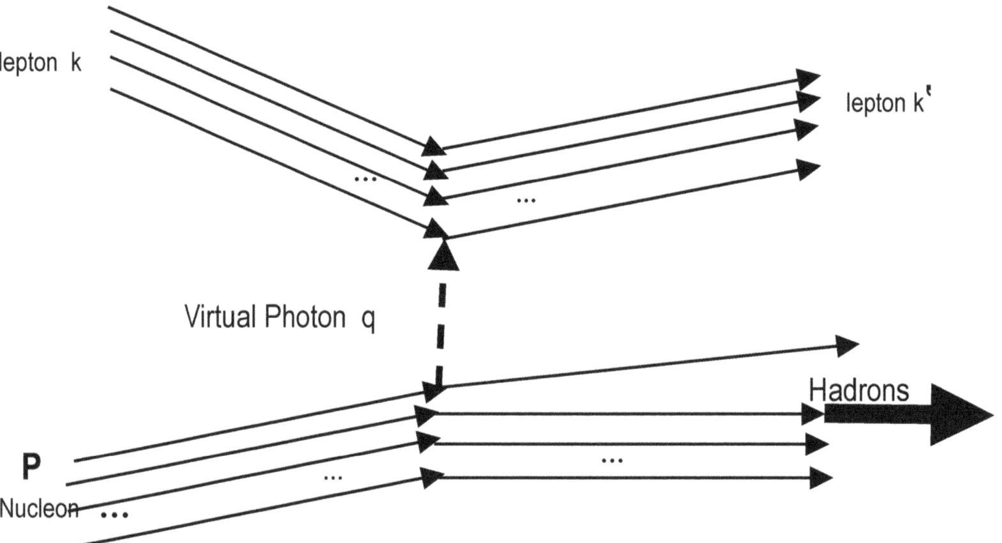

Figure 6.1. The kinematics of lepton-nucleon Deep Inelastic Scattering. The leptons, nucleons and virtual photon are all composed of partons of the Nth state.

---

[40] Eqs. 6.4 qnd 6.5 appear as eqs. 2.14 and 2.15 in Bjorken and Paschos.

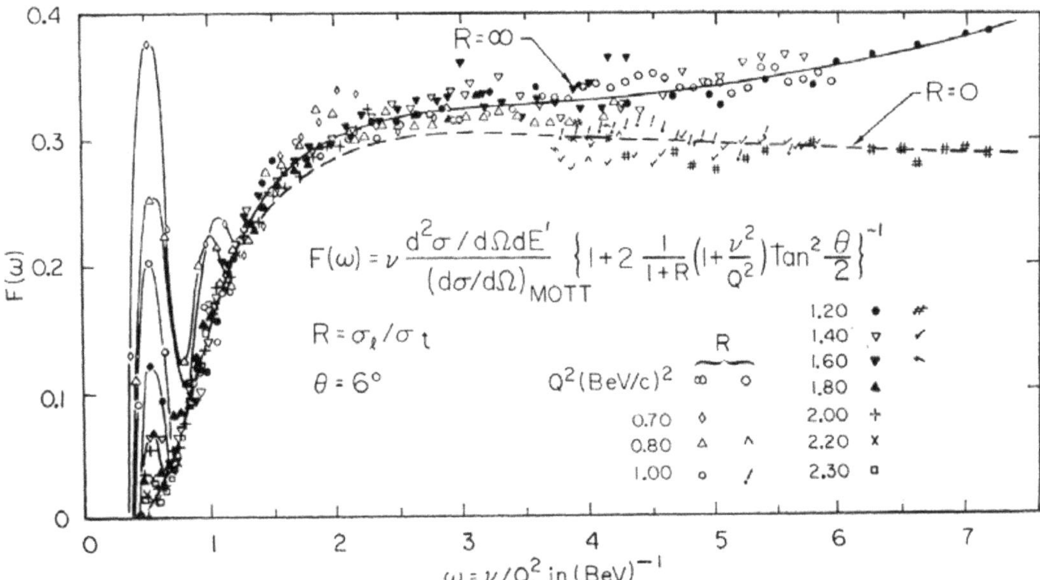

Figure 6.2. Plot[41] of F(ω) = νW₂(ω) vs. ω = 2M/x.

6.2 Gambol Model of Deep Inelastic Lepton-Nucleon Scattering

A proton consists of two u quarks and one d quark. Our gambol[42] model specifies gambol configurations for each fundamental particle. Our proton model thus has three gambol configurations – one configuration for each quark. We will assume that the configurations are approximately the same except for the total electromagnetic charges, which generate electric charges of 2/3, 2/3 and -1/3.

Following the approach of the preceding chapters we assume each quark is described by a set of fractionations where each fractionation "divides" the quark into s gambols where s = 1, 2, 4, 8, 16, … . (See eq. 4.1.) We label each fractionation with the integer

$$n = s \qquad (6.6)$$

We assume that there is a probability P(n) for each possible fractionation. Thus a quark is a "superposition" of fractionations. Later we will consider the probability of a fractionation n has the form similar to a Planck blackbody radiation distribution. We view the gambols as a form of quanta 'bouncing around" within a quark (or other fundamental particle).

---

[41] Rapporteur presentation of W. H. K. Panofsky in *Proceedings of the Fourteenth International Conference on High Energy Physics, Vienna, 1968* (CERN, Geneva, 1968) pp. 36-37 based on the research of E. Bloom *et al.* See Fig. 2 in Bjorken and Paschos referenced earlier. Figure courtesy of the American Physical Society.

[42] We chose to use gambol rather than parton since gambols have distinctively different features and since gambols will be used in the formulation of other particle interactions.

Concept

We view each hadron as made of a specific number of quarks (plus possibly a quark sea). A proton is made of three quarks that are confined within the proton by the Strong Interaction. Gambol theory assumes a fundamental particle may be viewed as a probabilistic set of fractionations of the particle into gambols which are physical but not "real" in the sense that there are a number of fractionations that follow a probability distribution. Gambols are thus a theoretical construct. We may view a fundamental particle as a probabilistic sum of fractionations into gambols—with each fractionation having a certain probability. In contrast a hadron is composed of a specific number of quarks that are "real" although they are confined within the hadron. Gambols are similarly confined but the gambol content of a particle varies with the fractionation—thus their "unreality." The gambol concept, if it applies to all fundamental particles, brings our understanding of the nature of matter to a deeper level. All particles are composed of gambols—a much "finer" substance of Reality. We see gambols $\rightarrow$ fundamental particles[43] $\rightarrow$ elementary particles $\rightarrow$ matter. The Gambol Model is supported by experimental evidence in Fig. 6.2.

We also believe that there is a probability distribution in gambol velocity for each fractionation. The velocities might have a Maxwell-Boltzmann distribution.

Based on these considerations we turn to the e-p structure function in our gambol formulation which we assume has the same form as eq. 6.5. We modify eq. 6.5 to

$$F(\omega) = vW_2(v, q^2) = \Sigma_n P(n) (\Sigma_i Q_i^2)_n f_n(x)$$

where n is the number of gambols and

$$
\begin{aligned}
&n = s \\
&p_{gambol} = xP \\
&x = 1/s \\
&\omega = 1/(2Mx) = s/(2M)
\end{aligned}
\tag{6.7}
$$

with M being the proton mass and P its longitudinal momentum in the infinite momentum frame. The fractionation s ranges from 1 to infinity in powers of two. We now extend the parameter s to be continuous. Consequently n is continuous as well.

We assume $F(\omega)$ is a probability distribution in $\omega = s/(2M)$. Each value of s, which corresponds to a fractionation into n = s gambols, determines the value of $F(\omega)$ for the value $\omega = s/(2M)$. Each value of ω specifies a specific fractionation s. Although s is defined as powers of two, we extend s to be a continuous parameter, just as ω is a continuous parameter.

We define $F_q(\omega)$ as an integral over fractionations for a quark

$$F_q(1/2M\omega) = F_q(x) = vW_{2q}(v, q^2) = \int dn\, \delta(n - x)\, V(n)f(n) \tag{6.8}$$
$$= V(x)f(x)$$

---

[43] The fundamental particles are quarks, leptons, and bosons. The elementary particles are hadrons, leptons, vector bosons and Higgs bosons.

with V(n) being a Planckian gambol fractionation probability distribution and f(n) being the Maxwell-Boltzmann gambol velocity distribution for fractionation n. We set f(n) = 1.

The function $F(\omega)$ is the probability of three n gambol states: one state for each of the three quarks, which we assume have approximately the same probability distribution:

$$F(x) = \Sigma_i\, Q_i^2\, F_q(x) = (4/9 + 4/9 + 1/9)\, F_q(\omega) = F_q(x) \qquad (6.9)$$

6.2.1 Calculation of the Probability of Fractionation V(ω)

We assume a set of gambols have a certain probability of fractionation into s gambols. We further assume that this probability is a Planck blackbody radiation distribution. See chapter 11 for a detailed derivation of the Planckian probability distribution. We view a fractionation of a particle into quasi-free gambols as occupying a confined space within a fundamental particle similar to photons within a blackbody.

The Planck distribution has the form:

$$u(\varepsilon) \equiv u(\nu) = 8\pi\varepsilon^3/h^2c^3/(e^{\varepsilon/kT} - 1)$$
$$= 8\pi h\nu^3/c^3\; 1/(e^{h\nu/kT} - 1) \qquad (6.10)$$

where ν is the frequency and

$$\varepsilon = \text{fractionation energy} = h\nu \qquad (6.11)$$

with h being Planck/s constant.

We normalize the integral of the probability distribution to one:

$$1 = \int_0^\infty d\varepsilon\; c/(4\sigma T^4)\, u(\nu) \qquad (6.12)$$

where

$$\sigma = 2\pi^5\, k^4/(15h^2c^2) \qquad (6.13)$$

We define the normalized Planck distribution:

$$U(\varepsilon) = 15/(\pi kT)^4\; \varepsilon^3/(e^{\varepsilon/kT} - 1) \qquad (6.14)$$

with normalization

$$1 = \int_0^\infty d\varepsilon\; U(\varepsilon) \qquad (6.15)$$

The Planck distribution was defined for massless photons. Particles have mass. The proton has mass M about 1 GeV/c^2. Gambols have a small mass m_{g0} which satisfies $m_{g0} = xM = M/s$ since the parton 4-momentum squared equals x^2 times the proton's 4-momentum squared. We therefore modify the energy ε to interpolate between the proton and gambol masses. Thus we now specify the *gambol distribution energy* at the lowest energy with nucleon velocity **v = 0**:

$$\varepsilon = \tfrac{1}{2}\, m_{g0} v^2 \;\rightarrow\; \varepsilon(s) = c^2(m_{g0}\, s + M)/(s+1) = (s m_{g0} + M)/(s+1) \qquad (6.16)$$

setting $c = 1$ where m_{g0} is the rest energy of a gambol. The gambol energy interpolates between $s = 0$ (with no fractionation[44] where $\varepsilon = M$ the proton mass), and $s = \infty$ (infinite fractionation where $\varepsilon = m_{g0}$ the mass of a gambol). Gambols are taken to be at rest for any fractionation.

$$s = 0 \qquad \varepsilon = Mc^2$$
$$s = \infty \qquad \varepsilon = m_{g0}c^2$$

We find the energy of a gambol in this approximation varies with the fractionation s.

Eq. 6.14 with ε given by eq. 6.16 is the Planckian probability distribution for the $n = s$ fractionation gambol state. We now define K with

$$kT = M/K \qquad (6.17)$$

resulting in the deep inelastic gambol Planckian distribution:[45]

$$U_g(\varepsilon(s)) = 15\,N\,(K/\pi M)^4\,\varepsilon^3/(e^{\varepsilon K/M} - 1) \qquad (6.18)$$

where N is a normalization factor for the photon – gambol transition with $\varepsilon = \varepsilon(s)$ given by eq. 6.16.

6.2.2 Deep Inelastic Probability $F(\omega)$

The gambol ω distribution of $F(\omega)$ is

$$F(\omega) = U_g(\varepsilon(2M\omega)) \qquad (6.19)$$

with $s = 2M\omega$.

Eq. 6.19 is plotted in Fig. 6.3 vs. ω with the parameters

$$M = 1.0 \text{ GeV}/c^2 \qquad\qquad (6.20)$$
$$m_{g0}{}^{46} = M/8 = 0.125 \text{ GeV}/c^2$$
$$kT^{47} = 0.328\, M/m_b = 0.0785 \text{ GeV}/c^2$$
$$K = M/\,kT = 12.74$$
$$N = 0.122$$

It corresponds to Fig. 6.2. The plots in Figs. 6.2, 6.3 and 6.4 are a close match and suggest that the gambol model of deep inelastic proton scattering may be valid. The

---

[44] The $s = 0$ case is based on an extrapolation of the fractionation parameter to the range $s = 1$ to $s = \infty$.
[45] We generalize this definition in chapter 9.
[46] We chose this value for the effective gambol mass. The plot of $F(\omega)$ has a corresponding peak at $s = 8$.
[47] This expression for kT has the same form as kT expressions for quarks. See section 9.8. The temperature quantity is that of the proton gambols.

calculation of the gambol deep inelastic neutron scattering $F(\omega)$ would appear to give similar results. Spin dependent gambol deep inelastic scattering remains to be done.

The value of m_{g0} corresponds to $x = 0.125$. In terms of x the ε of eq. 6.16 is

$$\varepsilon = (m_{g0} + xM)/(1 + x) \qquad (6.21)$$

The effective gambol mass is $m_{g0} = 0.125M = 125$ MeV/c^2, which is 1/8 of the proton mass. The corresponding fractionation value is thus

$$S_{\text{effective}} = 8 \qquad (6.22)$$

The appearance of *effective* fractionation 1/8 may be regarded as strongly supportive of fractionation and Limos.

The effective value of kT is 78.5 MeV/c^2. Note $m_{g0} = 3/2\ kT$. The normalization is

$$N \cong 3/(2K) \cong m_{g0}/M \qquad (6.23)$$

The above calculation is approximate. The parameter values are best estimates. They are extremely sensitive to small changes.[48]

The excellence of the gambol results in comparison to experiment supports gambol fractionation as a physically meaningful approach. It also supports Limos and Cosmos Theory within which it resides.

The approximations of this calculation are:

1. We assume the probability of fractionation is proportional to the Planck distribution suitably changed to reflect non-zero gambol masses.

2. We use of proton M instead of gambol m_{g0} for the x and ω parameters.

3. We use an interpolation for the effective gambol mass.

4. We assume the gambols have mass and are at rest. We use the lowest energy gambol energy with $v = c = 1$.

The Planck distribution modified for massive gambols clearly gives an accurate description of the deep inelastic e-proton $F(\omega)$ and supports the Limos sector of Cosmos Theory. See Fig. 6.5 for a numerical comparison.

6.3 Gambol Deep Inelastic Model

The calculations of the Gambol Model are based on the following points:

---

[48] The calculation was performed in Excel in double precision.

1. An S-matrix factor due to the gambol-photon interaction using the proton spin and mass in the S-matrix factor together with the gambol charge and mass m_{g0}. The imaginary part of this factor leads to a W_2 factor in eq. 6.4 above

$$\delta((q + p_g)^2 - m_{g0}{}^2) \;\rightarrow\; \delta(\nu - Q^2/(2xM)) \qquad (6.24)$$

after extracting a $1/(2xM)$ factor from the δ-function where p_g is the gambol four-momentum.

2. A Planckiam probability distribution based on an energy that interpolates between the proton mass and the gambol mass m_{g0} with gambols assumed to be at rest.[49]

3. An integration over the fractionation parameter $s = 1/x$ resulting in a W_2 expression similar to eq. 6.5.

4. The calculation may be improved with some mild parameter changes if experiments yield a detailed $F(\omega)$ curve. The calculation assumes the gambols are at rest in the proton's reference frame. If the gambols have a velocity, then a Q^2 dependence would be introduced as recent experiments show.

The Planckian probability distribution may be placed in the form of an S-matrix-like term:[50]

$$S_g = \sqrt{U_g(\varepsilon_g)}\; e^{i\varphi} \qquad (6.25)$$

where φ is an arbitrary phase with

$$S_g{}^\dagger S_g = U_g(\varepsilon_g) \qquad (6.26)$$

becoming the probability factor in eq. 6.18 after an integration that transforms the $s = 1/x$ parameter into $s = 2M\omega$.

Thus we view the deep inelastic S-matrix as composed of

$$S = S_g S_{\text{e-gproton}} \qquad (6.27)$$

with S_g given by eq. 6.25 and $S_{\text{e-gproton}}$ being in the e-proton S-matrix element modified to the gambols as in point 1 above: gambol mass m_{g0}, and momentum.

---

[49] We specify the gambols at each fractionation value to be at rest. The possibility presents itself of gambol configurations distributed within a particle in the spatial form of symmetric configurations of the type of crystallographic point groups S_n where $n = 2^s$. We may view the spatial pattern of a fractionated particle as a regular polygonal pattern of gambol points distributed on a sphere near the surface of the particle to minimize gambol interactions. A particle then appears to be a crystal.

[50] S_g could be viewed as resulting from a simple harmonic oscillator model in which the number operator states may be superimposed to have a form leading to the Planckian probability distribution.

The form of $W_2(v, q^2)$ in eq. 6.4 may be re-expressed as

$$W_2(v, q^2) = \int ds\, U_g(\varepsilon_g(s))\, (\Sigma_i\, Q_i^2)\, \delta(v - sQ^2/(2M)) \qquad (6.28)$$

which, after integration, gives

$$F(\omega) = vW_2(v, q^2) = (\Sigma_i\, Q_i^2)\, U_g(\varepsilon_g(s = 2Mv/Q^2 = 2M\omega)) \qquad (6.29)$$

$$= (\Sigma_i\, Q_i^2)\, U_g(\varepsilon_g(2M\omega)$$

It is important to note that the gambols, for all fractionations s, are treated as massive particles at rest, with no interactions in this Deep Inelastic Gambol Model.

The deep inelastic e-proton experiments give an instantaneous view of the form of the proton and the quarks within it. These particles have a structure dictated by the strong interactions within the proton as evidenced by the form of the $F(\omega)$ curve in Fig. 6.2.

The gambol fractionations of the proton use massive, stationary (rest frame), non-interacting gambols. Thus the gambol-based approach offers a new interaction-free view of proton structure. In the following chapters we will consider hadron-hadron scattering, and particle decay within the gambol fractionation framework. The success of a gambol-based approach offers a new "handle" for the understanding of particle dynamics. *Complex interaction-based scattering is transformed to fractionated "free" gambol dynamics.*

The plots in Figs. 6.3 and 6.4 have a maximum at $s = 2M\omega = 8$ indicating that the proton's optimal number of gambols is 8.

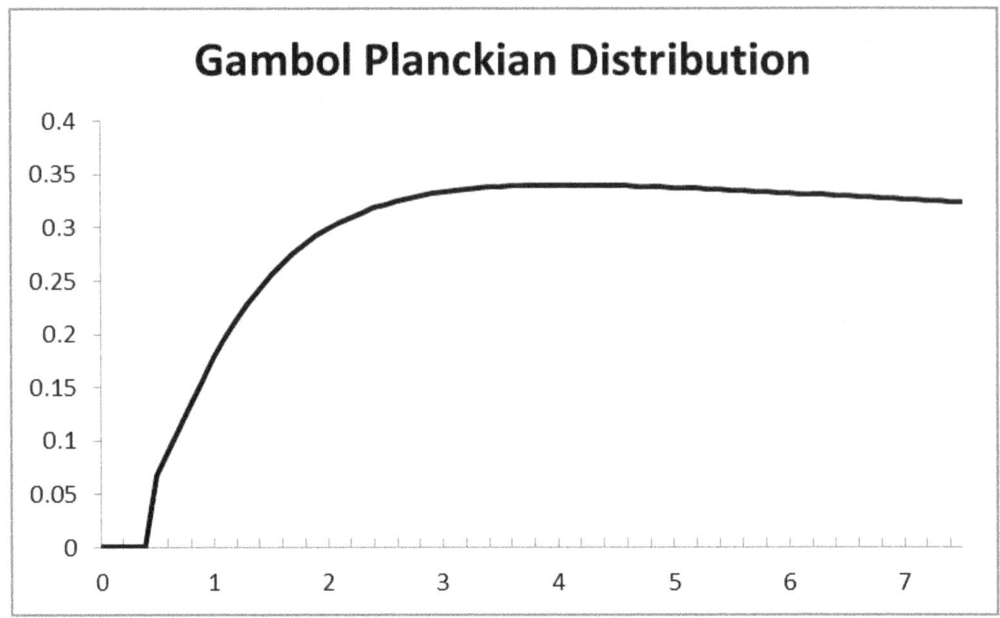

Figure 6.3. The gambol Planckian distribution of eq. 6.19 plotted against ω for the R = 0 proton deep inelastic scattering.

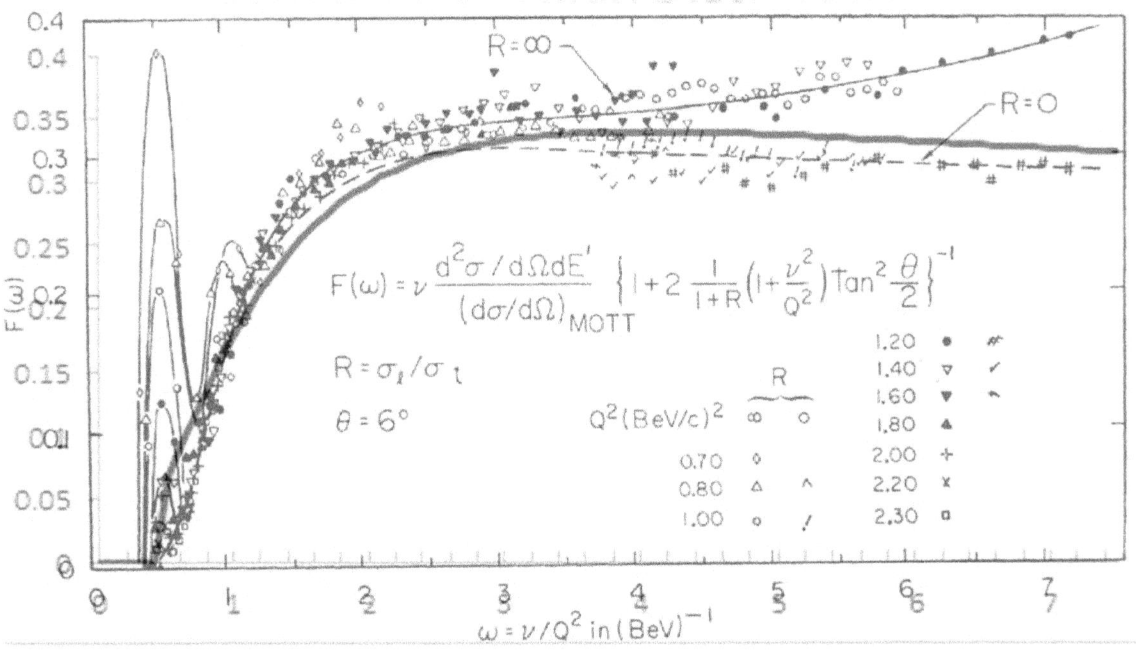

$$F(\omega) = \nu \frac{d^2\sigma/d\Omega dE'}{(d\sigma/d\Omega)_{MOTT}}\left\{1 + 2\frac{1}{1+R}\left(1+\frac{\nu^2}{Q^2}\right)Tan^2\frac{\theta}{2}\right\}^{-1}$$

Figure 6.4. Plot of F(ω) = vW₂(ω) vs. ω = 2M/x with the gambol Planckian distribution of eq. 6.19 superimposed on it. Match! *Note that the experimental values have error bars so a precise match is not possible. The curve may be adjusted by slightly changing some parameters.*

ω	1	2	3	4	5	6	7
$F(\omega)_{experiment}$[51]	0.19	0.3	0.3	0.3	0.3	0.3	0.3
$F(\omega)_{Planckian}$	0.18	0.299	0.33	0.34	0.34	0.33	0.326

Figure 6.5. Table of Gambol Planckian distribution values vs. experiment (noting some spread in experimental values).

---

[51] Approximately estimated from Fig. 6.2.

13. Structure of Physical Cosmos Spaces and Universes

Cosmos Theory has led us to a new, deeper view of matter, interactions and space-times. The theory develops universes from the Cosmos spectrum of spaces using a Cosmos Universe Consistency condition that fixes the maximum dimension of Physical Cosmos spaces to 18. We can view the development pattern of Cosmos Theory as:

1. Cosmos Spaces
2. Consistency Condition for maximum dimension space
3. Parent Universe(s)
4. Children, Grandchildren, … Universes
5. Our UST Universe
6. Universe Particles
7. Universe Interactions

13.1 Unifying Principle for the Sequences

The seven sequences that appear: Cosmos spaces, Coupling Constants, Gravitation coupling constant, Up-Type fermion masses, Down-Type fermion masses, Vector Boson W and Z masses, and universe masses are all based on Mathematics and Geometry:

1. Cosmos spaces based on Mathematics of totally asymmetric tensors (geometry of surfaces and volumes) and fermion creation and annihilation operators.

2. Coupling Constants based geometry and spin of fermion creation and annihilation operators.

3. Masses based on geometry and the Cosmos symmetry group of the UST layer.

4. Universe masses based on geometry.

5. Vector boson: W and Z masses based on coupling constants.

We conclude the total Lagrangian for our universe is partly fixed:

1. Cosmos Theory gives spaces and dimension arrays, coupling constants and masses.

2. The UST Lagrangian is almost totally defined by using Cosmos Theory interactions, coupling constants and masses (and scalar and vector bosons); and by Yang-Mills formulations.

The constants in the total elementary particle Lagrangian are thus known except for the possible role of Higgs bosons.

Possible additional particles and interactions remain to be found. The remaining great issue is the determination of scattering matrix elements.

The above sequences of spaces, coupling constants, masses, and universe masses all reflect a fundamental basis in the features of dimensions in Geometry emanating from totally antisymmetric tensors through particles with internal universe shells.

REFERENCES

Akhiezer, N. I., Frink, A. H. (tr), 1962, *The Calculus of Variations* (Blaisdell Publishing, New York, 1962).

Bjorken, J. D., Drell, S. D., 1964, *Relativistic Quantum Mechanics* (McGraw-Hill, New York, 1965).

Bjorken, J. D., Drell, S. D., 1965, *Relativistic Quantum Fields* (McGraw-Hill, New York, 1965).

Blaha, S., 1995, *C++ for Professional Programming* (International Thomson Publishing, Boston, 1995).

________, 1998, *Cosmos and Consciousness* (Pingree-Hill Publishing, Auburn, NH, 1998 and 2002).

________, 2002, *A Finite Unified Quantum Field Theory of the Elementary Particle Standard Model and Quantum Gravity Based on New Quantum Dimensions™ & a New Paradigm in the Calculus of Variations* (Pingree-Hill Publishing, Auburn, NH, 2002).

________, 2004, *Quantum Big Bang Cosmology: Complex Space-time General Relativity, Quantum Coordinates™ Dodecahedral Universe, Inflation, and New Spin 0, ½, 1 & 2 Tachyons & Imagyons* (Pingree-Hill Publishing, Auburn, NH, 2004).

______, 2005a, *Quantum Theory of the Third Kind: A New Type of Divergence-free Quantum Field Theory Supporting a Unified Standard Model of Elementary Particles and Quantum Gravity based on a New Method in the Calculus of Variations* (Pingree-Hill Publishing, Auburn, NH, 2005).

______, 2005b, *The Metatheory of Physics Theories, and the Theory of Everything as a Quantum Computer Language* (Pingree-Hill Publishing, Auburn, NH, 2005).

______, 2005c, *The Equivalence of Elementary Particle Theories and Computer Languages: Quantum Computers, Turing Machines, Standard Model, Superstring Theory, and a Proof that Gödel's Theorem Implies Nature Must Be Quantum* (Pingree-Hill Publishing, Auburn, NH, 2005).

______, 2006a, *The Foundation of the Forces of Nature* (Pingree-Hill Publishing, Auburn, NH, 2006).

______, 2006b, *A Derivation of ElectroWeak Theory based on an Extension of Special Relativity; Black Hole Tachyons; & Tachyons of Any Spin.* (Pingree-Hill Publishing, Auburn, NH, 2006).

______, 2007a, *Physics Beyond the Light Barrier: The Source of Parity Violation, Tachyons, and A Derivation of Standard Model Features* (Pingree-Hill Publishing, Auburn, NH, 2007).

______, 2007b, *The Origin of the Standard Model: The Genesis of Four Quark and Lepton Species, Parity Violation, the ElectroWeak Sector, Color SU(3), Three Visible Generations of Fermions, and One Generation of Dark Matter with Dark Energy* (Pingree-Hill Publishing, Auburn, NH, 2007).

______, 2008a, *A Direct Derivation of the Form of the Standard Model From GL(16) (Pingree-Hill Publishing, Auburn, NH, 2008).*

______, 2008b, *A Complete Derivation of the Form of the Standard Model With a New Method to Generate Particle Masses Second Edition* (Pingree-Hill Publishing, Auburn, NH, 2008)

______, 2009, *The Algebra of Thought & Reality: The Mathematical Basis for Plato's Theory of Ideas, and Reality Extended to Include A Priori Observers and Space-Time Second Edition* (Pingree-Hill Publishing, Auburn, NH, 2009).

______, 2010a, *Operator Metaphysics: A New Metaphysics Based on a New Operator Logic and a New Quantum Operator Logic that Lead to a Mathematical Basis for Plato's Theory of Ideas and Reality* (Pingree-Hill Publishing, Auburn, NH, 2010).

______, 2010b, *The Standard Model's Form Derived from Operator Logic, Superluminal Transformations and GL(16)* (Pingree-Hill Publishing, Auburn, NH, 2010).

______, 2010c, *SuperCivilizations: Civilizations as Superorganisms* (McMann-Fisher Publishing, Auburn, NH, 2010).

______, 2011a, *21st Century Natural Philosophy Of Ultimate Physical Reality* (McMann-Fisher Publishing, Auburn, NH, 2011).

______, 2011b, *All the Universe! Faster Than Light Tachyon Quark Starships & Particle Accelerators with the LHC as a Prototype Starship Drive Scientific Edition* (Pingree-Hill Publishing, Auburn, NH, 2011).

______, 2011c, *From Asynchronous Logic to The Standard Model to Superflight to the Stars* (Blaha Research, Auburn, NH, 2011).

______, 2012a, *From Asynchronous Logic to The Standard Model to Superflight to the Stars volume 2: Superluminal CP and CPT, U(4) Complex General Relativity and The Standard Model, Complex Vierbein General Relativity, Kinetic Theory, Thermodynamics* (Blaha Research, Auburn, NH, 2012).

______, 2012b, *Standard Model Symmetries, And Four And Sixteen Dimension Complex Relativity; The Origin Of Higgs Mass Terms* (Blaha Reasearch, Auburn, NH, 2012).

______, 2013a, *Multi-Stage Space Guns, Micro-Pulse Nuclear Rockets, and Faster-Than-Light Quark-Gluon Ion Drive Starships* (Blaha Research, Auburn, NH, 2013).

______, 2013b, *The Bridge to Dark Matter; A New Sibling Universe; Dark Energy; Inflatons; Quantum Big Bang; Superluminal Physics; An Extended Standard Model Based on Geometry* (Blaha Reasearch, Auburn, NH, 2013).

______, 2014a, *Universes and Megaverses: From a New Standard Model to a Physical Megaverse; The Big Bang; Our Sibling Universe's Wormhole; Origin of the Cosmological Constant, Spatial Asymmetry of the Universe, and its Web of Galaxies; A Baryonic Field between Universes and Particles; Megaverse Extended Wheeler-DeWitt Equation* (Blaha Reasearch, Auburn, NH, 2014).

______, 2014b, *All the Megaverse! Starships Exploring the Endless Universes of the Cosmos Using the Baryonic Force* (Blaha Research, Auburn, NH, 2014).

______, 2014c, *All the Megaverse! II Between Megaverse Universes: Quantum Entanglement Explained by the Megaverse Coherent Baryonic Radiation Devices – PHASERs Neutron Star Megaverse Slingshot Dynamics Spiritual and UFO Events, and the Megaverse Microscopic Entry into the Megaverse* (Blaha Research, Auburn, NH, 2014).

______, 2015a, *PHYSICS IS LOGIC PAINTED ON THE VOID: Origin of Bare Masses and The Standard Model in Logic, U(4) Origin of the Generations, Normal and Dark Baryonic Forces, Dark Matter, Dark Energy, The Big Bang, Complex General Relativity, A Megaverse of Universe Particles* (Blaha Research, Auburn, NH, 2015).

______, 2015b, *PHYSICS IS LOGIC Part II: The Theory of Everything, The Megaverse Theory of Everything, U(4)⊗U(4) Grand Unified Theory (GUT), Inertial Mass = Gravitational Mass, Unified Extended Standard Model and a New Complex General Relativity with Higgs Particles, Generation Group Higgs Particles* (Blaha Research, Auburn, NH, 2015).

______, 2015c, *The Origin of Higgs ("God") Particles and the Higgs Mechanism: Physics is Logic III, Beyond Higgs – A Revamped Theory With a Local Arrow of Time, The Theory of Everything Enhanced, Why Inertial Frames are Special, Universes of the Mind* (Blaha Research, Auburn, NH, 2015).

______, 2015d, *The Origin of the Eight Coupling Constants of The Theory of Everything: U(8) Grand Unified Theory of Everything (GUTE), S^8 Coupling Constant Symmetry, Space-Time Dependent Coupling Constants, Big Bang Vacuum Coupling Constants, Physics is Logic IV* (Blaha Research, Auburn, NH, 2015).

______, 2016a, *New Types of Dark Matter, Big Bang Equipartition, and A New U(4) Symmetry in the Theory of Everything: Equipartition Principle for Fermions, Matter is 83.33% Dark, Penetrating the Veil of the Big Bang, Explicit QFT Quark Confinement and Charmonium, Physics is Logic V* (Blaha Research, Auburn, NH, 2016).

______, 2016b, *The Periodic Table of the 192 Quarks and Leptons in The Theory of Everything: The U(4) Layer Group, Physics is Logic VI* (Blaha Research, Auburn, NH, 2016).

______, 2016c, *New Boson Quantum Field Theory, Dark Matter Dynamics, Dark Matter Fermion Layer Mixing, Genesis of Higgs Particles, New Layer Higgs Masses, Higgs Coupling Constants, Non-Abelian Higgs Gauge Fields, Physics is Logic VII* (Blaha Research, Auburn, NH, 2016).

______, 2016d, *Unification of the Strong Interactions and Gravitation: Quark Confinement Linked to Modified Short-Distance Gravity; Physics is Logic VIII* (Blaha Research, Auburn, NH, 2016).

______, 2016e, *MoND: Unification of the Strong Interactions and Gravitation II, Quark Confinement Linked to Large-Scale Gravity, Physics is Logic IX* (Blaha Research, Auburn, NH, 2016).

______, 2016f, *CQ Mechanics: A Unification of Quantum & Classical Mechanics, Quantum/Semi-Classical Entanglement, Quantum/Classical Path Integrals, Quantum/Classical Chaos* (Blaha Research, Auburn, NH, 2016).

______, 2016g, *GEMS Unified Gravity, ElectroMagnetic and Strong Interactions: Manifest Quark Confinement, A Solution for the Proton Spin Puzzle, Modified Gravity on the Galactic Scale* (Pingree Hill Publishing, Auburn, NH, 2016).

______, 2016h, *Unification of the Seven Boson Interactions based on the Riemann-Christoffel Curvature Tensor* (Pingree Hill Publishing, Auburn, NH, 2016).

______, 2017a, *Unification of the Eleven Boson Interactions based on 'Rotations of Interactions'* (Pingree Hill Publishing, Auburn, NH, 2017).

______, 2017b, *The Origin of Fermions and Bosons, and Their Unification* (Pingree Hill Publishing, Auburn, NH, 2017).

______, 2017c, *Megaverse: The Universe of Universes* (Pingree Hill Publishing, Auburn, NH, 2017).

______, 2017d, *SuperSymmetry and the Unified SuperStandard Model* (Pingree Hill Publishing, Auburn, NH, 2017).

______, 2017e, *From Qubits to the Unified SuperStandard Model with Embedded SuperStrings: A Derivation* (Pingree Hill Publishing, Auburn, NH, 2017).

______, 2017f, *The Unified SuperStandard Model in Our Universe and the Megaverse: Quarks, … ,* (Pingree Hill Publishing, Auburn, NH, 2017).

______, 2018a, *The Unified SuperStandard Model and the Megaverse SECOND EDITION A Deeper Theory based on a New Particle Functional Space that Explicates Quantum Entanglement Spookiness (Volume 1)* (Pingree Hill Publishing, Auburn, NH, 2018).

______, 2018b, *Cosmos Creation: The Unified SuperStandard Model, Volume 2, SECOND EDITION* (Pingree Hill Publishing, Auburn, NH, 2018).

______, 2018c, *God Theory* (Pingree Hill Publishing, Auburn, NH, 2018).

______, 2018d, *Immortal Eye: God Theory: Second Edition* (Pingree Hill Publishing, Auburn, NH, 2018).

______, 2018e, *Unification of God Theory and Unified SuperStandard Model THIRD EDITION* (Pingree Hill Publishing, Auburn, NH, 2018).

______, 2019a, *Calculation of: QED α = 1/137, and Other Coupling Constants of the Unified SuperStandard Theory* (Pingree Hill Publishing, Auburn, NH, 2019).

______, 2019b, *Coupling Constants of the Unified SuperStandard Theory SECOND EDITION* (Pingree Hill Publishing, Auburn, NH, 2019).

______, 2019c, *New Hybrid Quantum Big_Bang–Megaverse_Driven Universe with a Finite Big Bang and an Increasing Hubble Constant* (Pingree Hill Publishing, Auburn, NH, 2019).

______, 2019d, *The Universe, The Electron and The Vacuum* (Pingree Hill Publishing, Auburn, NH, 2019).

______, 2019e, *Quantum Big Bang – Quantum Vacuum Universes (Particles)* (Pingree Hill Publishing, Auburn, NH, 2019).

______, 2019f, *The Exact QED Calculation of the Fine Structure Constant Implies ALL 4D Universes have the Same Physics/Life Prospects* (Pingree Hill Publishing, Auburn, NH, 2019).

______, 2019g, *Unified SuperStandard Theory and the SuperUniverse Model: The Foundation of Science* (Pingree Hill Publishing, Auburn, NH, 2019).

______, 2020a, *Quaternion Unified SuperStandard Theory (The QUeST) and Megaverse Octonion SuperStandard Theory (MOST)* (Pingree Hill Publishing, Auburn, NH, 2020).

______, 2020b, *United Universes Quaternion Universe - Octonion Megaverse* (Pingree Hill Publishing, Auburn, NH, 2020).

______, 2020c, *Unified SuperStandard Theories for Quaternion Universes & The Octonion Megaverse* (Pingree Hill Publishing, Auburn, NH, 2020).

______, 2020d, *The Essence of Eternity: Quaternion & Octonion SuperStandard Theories* (Pingree Hill Publishing, Auburn, NH, 2020).

______, 2020e, *The Essence of Eternity II* (Pingree Hill Publishing, Auburn, NH, 2020).

______, 2020f, *A Very Conscious Universe* (Pingree Hill Publishing, Auburn, NH, 2020).

______, 2020g, *Hypercomplex Universe* (Pingree Hill Publishing, Auburn, NH, 2020).

______, 2020h, *Beneath the Quaternion Universe* (Pingree Hill Publishing, Auburn, NH, 2020).

______, 2020i, *Why is the Universe Real? From Quaternion & Octonion to Real Coordinates* (Pingree Hill Publishing, Auburn, NH, 2020).

______, 2020j, *The Origin of Universes: of Quaternion Unified SuperStandard Theory (QUeST); and of the Octonion Megaverse (UTMOST)* (Pingree Hill Publishing, Auburn, NH, 2020).

______, 2020k, *The Seven Spaces of Creation: Octonion Cosmology* (Pingree Hill Publishing, Auburn, NH, 2020).
______, 2020l, *From Octonion Cosmology to the Unified SuperStandard Theory of Particles* (Pingree Hill Publishing, Auburn, NH, 2020).

______, 2021a, *Pioneering the Cosmos* (Pingree Hill Publishing, Auburn, NH, 2021).

______, 2021b, *Pioneering the Cosmos II* (Pingree Hill Publishing, Auburn, NH, 2021).

______, 2021c, *Beyond Octonion Cosmology* (Pingree Hill Publishing, Auburn, NH, 2021).

______, 2021d, *Universes are Particles* (Pingree Hill Publishing, Auburn, NH, 2021).

______, 2021e, *Octonion-like dna-based life, Universe expansion is decay, Emerging New Physics* (Pingree Hill Publishing, Auburn, NH, 2021).

______, 2021f, *The Science of Creation New Quantum Field Theory of Spaces* (Pingree Hill Publishing, Auburn, NH, 2021).

______, 2021g, *Quantum Space Theory With Application to Octonion Cosmology & Possibly To Fermionic Condensed Matter* (Pingree Hill Publishing, Auburn, NH, 2021).

______, 2021h, *21st Century Natural Philosophy of Octonion Cosmology , and Predestination, Fate, and Free Will* (Pingree Hill Publishing, Auburn, NH, 2021).

______, 2021i, *Beyond Octonion Cosmology II : Origin of the Quantum; A New Generalized Field Theory (GiFT); A Proof of the Spectrum of Universes; Atoms in Higher Universes* (Pingree Hill Publishing, Auburn, NH, 2021).

______, 2021j, *Integration of General Relativity and Quantum Theory: Octonion Cosmology, GiFT, Creation/Annihilation Spaces CASe, Reduction of Spaces to a Few Fermions and Symmetries in Fundamental Frames* (Pingree Hill Publishing, Auburn, NH, 2021).

______, 2022a, *New View of Octonion Cosmology Based on the Unification of General Relativity and Quantum Theory* (Pingree Hill Publishing, Auburn, NH, 2022).

______, 2022b, *The Dust Beneath Hypercomplex Cosmology* (Pingree Hill Publishing, Auburn, NH, 2022).

______, 2022c, *Passing Through Nature to Eternity: ProtoCosmos, HyperCosmos, Unified SuperStandard Theory* (Pingree Hill Publishing, Auburn, NH, 2022).

______, 2022d, *HyperCosmos Fractionation and Fundamental Reference Frame Based Unification: Particle Inner Space Basis of Parton and Dual Resonance Models* (Pingree Hill Publishing, Auburn, NH, 2022).

______, 2022e, *A New UniDimension ProtoCosmos and SuperString F-Theory Relation to the HyperCosmos* (Pingree Hill Publishing, Auburn, NH, 2022).

______, 2022f, *The Cosmic Panorama: ProtoCosmos, HyperCosmos,Unified SuperStandard Theory (UST) Derivation* (Pingree Hill Publishing, Auburn, NH, 2022).

______, 2022g, *Ultimate Origin: ProtoCosmos and HyperCosmos* (Pingree Hill Publishing, Auburn, NH, 2022).

______, 2023a, *UltraUnification and the Generation of the Cosmos* (Pingree Hill Publishing, Auburn, NH, 2023).

______, 2023b, *God and and Cosmos Theory* (Pingree Hill Publishing, Auburn, NH, 2023).

______, 2023c, *A New Completely Geometric SU(8) Cosmos Theory; New PseudoFermion Fields; Fibonacci-like Dimension Arrays; Ramsey Number Approximation* (Pingree Hill Publishing, Auburn, NH, 2023).

______, 2023d, *Newton's Apple is Now the Fermion* (Pingree Hill Publishing, Auburn, NH, 2023).

______, 2023e,*Cosmos Theory: The Sub-Particle Gambol Model* (Pingree Hill Publishing, Auburn, NH, 2023).

______, 2024a, *Cosmos-Universe-Particle-Gambol Theory* (Pingree Hill Publishing, Auburn, NH, 2024).

______, 2024b, *Fractal Cosmos Theory* (Pingree Hill Publishing, Auburn, NH, 2024).

______, 2024c, *Fractal Cosmic Curve: Tensor-Based CosmosTheory* (Pingree Hill Publishing, Auburn, NH, 2024).

                        REFERENCES

_____, 2024d, *The Eternal Form of Cosmos Theory* (Pingree Hill Publishing, Auburn, NH, 2024).

_____, 2024e, *The Eternal Form of Cosmos Theory Third Edition* (Pingree Hill Publishing, Auburn, NH, 2024).

_____, 2024f, *Fundamental Constants of Cosmos Theory and The Standard Model* (Pingree Hill Publishing, Auburn, NH, 2024).

_____, 2024g, *Quark, Lepton, W and Z Masses of Cosmos Theory and The Standard Model* (Pingree Hill Publishing, Auburn, NH, 2024).

_____, 2024h, *Geometric Cosmos Geometric Universe* (Pingree Hill Publishing, Auburn, NH, 2024).

Eddington, A. S., 1952, *The Mathematical Theory of Relativity* (Cambridge University Press, Cambridge, U.K., 1952).

Fant, Karl M., 2005, *Logically Determined Design: Clockless System Design With NULL Convention Logic* (John Wiley and Sons, Hoboken, NJ, 2005).

Feinberg, G. and Shapiro, R., 1980, *Life Beyond Earth: The Intelligent Earthlings Guide to Life in the Universe* (William Morrow and Company, New York, 1980).

Gelfand, I. M., Fomin, S. V., Silverman, R. A. (tr), 2000, *Calculus of Variations* (Dover Publications, Mineola, NY, 2000).

Giaquinta, M., Modica, G., Souchek, J., 1998, *Cartesian Coordinates in the Calculus of Variations* Volumes I and II (Springer-Verlag, New York, 1998).

Giaquinta, M., Hildebrandt, S., 1996, *Calculus of Variations* Volumes I and II (Springer-Verlag, New York, 1996).

Gradshteyn, I. S. and Ryzhik, I. M., 1965, *Table of Integrals, Series, and Products* (Academic Press, New York, 1965).

Heitler, W., 1954, *The Quantum Theory of Radiation* (Claendon Press, Oxford, UK, 1954).

Huang, Kerson, 1992, *Quarks, Leptons & Gauge Fields 2nd Edition* (World Scientific Publishing Company, Singapore, 1992).

Jost, J., Li-Jost, X., 1998, *Calculus of Variations* (Cambridge University Press, New York, 1998).

Kaku, Michio, 1993, *Quantum Field Theory*, (Oxford University Press, New York, 1993).

Kirk, G. S. and Raven, J. E., 1962, *The Presocratic Philosophers* (Cambridge University Press, New York, 1962).

Landau, L. D. and Lifshitz, E. M., 1987, *Fluid Mechanics 2nd Edition*, (Pergamon Press, Elmsford, NY, 1987).

Rescher, N., 1967, *The Philosophy of Leibniz* (Prentice-Hall, Englewood Cliffs, NJ, 1967).

Riesz, Frigyes and Sz.-Nagy, Béla, 1990, *Functional Analysis* (Dover Publications, New York, 1990).

Sakurai, J. J., 1964, *Invariance Principles and Elementary Particles* (Princeton University Press, Princeton, NJ, 1964).

Weinberg, S., 1972, *Gravitation and Cosmology* (John Wiley and Sons, New York, 1972).

Weinberg, S., 1995, *The Quantum Theory of Fields Volume I* (Cambridge University Press, New York, 1995).

INDEX

About the Author

Stephen Blaha is a well-known Physicist and Man of Letters with interests in Science, Society and civilization, the Arts, and Technology. He had an Alfred P. Sloan Foundation scholarship in college. He received his Ph.D. in Physics from Rockefeller University. He has served on the faculties of several major universities. He was also a Member of the Technical Staff at Bell Laboratories, a manager at the Boston Globe Newspaper, a Director at Wang Laboratories, and President of Blaha Software Inc. and of Janus Associates Inc. (NH).

Among other achievements he was a co-discoverer of the "r potential" for heavy quark binding developing the first (and still the only demonstrable) non-Aeolian gauge theory with an "r" potential; first suggested the existence of topological structures in superfluid He-3; first proposed Yang-Mills theories would appear in condensed matter phenomena with non-scalar order parameters; first developed a grammar-based formalism for quantum computers and applied it to elementary particle theories; first developed a new form of quantum field theory without divergences (thus solving a major 60 year old problem that enabled a unified theory of the Standard Model and Quantum Gravity without divergences to be developed); first developed a formulation of complex General Relativity based on analytic continuation from real space-time; first developed a generalized non-homogeneous Robertson-Walker metric that enabled a quantum theory of the Big Bang to be developed without singularities at $t = 0$; first generalized Cauchy's theorem and Gauss' theorem to complex, curved multi-dimensional spaces; received Honorable Mention in the Gravity Research Foundation Essay Competition in 1978; first developed a physically acceptable theory of faster-than-light particles; first derived a composition of extremums method in the Calculus of Variations; first quantitatively suggested that inflationary periods in the history of the universe were not needed; first proved Gödel's Theorem implies Nature must be quantum; provided a new alternative to the Higgs Mechanism, and Higgs particles, to generate masses; first showed how to resolve logical paradoxes including Gödel's Undecidability Theorem by developing Operator Logic and Quantum Operator Logic; first developed a quantitative harmonic oscillator-like model of the life cycle, and interactions, of civilizations; first showed how equations describing superorganisms also apply to civilizations. A recent book shows his theory applies successfully to the past 14 years of history and to *new* archaeological data on Andean and Mayan civilizations as well as Early Anatolian and Egyptian civilizations.

He first developed an axiomatic derivation of the form of The Standard Model from geometry – space-time properties – The Unified SuperStandard Model. It unifies all the known forces of Nature. It also has a Dark Matter sector that includes a Dark ElectroWeak sector with Dark doublets and Dark gauge interactions. It uses quantum coordinates to remove infinities that crop up in most interacting quantum field theories and additionally to remove the infinities that appear in the Big Bang and generate inflationary growth of the universe. It shows gravity has a MOND-like form without sacrificing Newton's Laws. It relates the interactions of the MOND-like sector of gravity with the r-potential of Quark Confinement. The axioms of the theory lead to the question of their origin. We suggest in the preceding edition of this book it can be attributed to an entity with God-like properties. We explore these properties in "God Theory" and show they predict that the Cosmos exists forever although individual universes (or incarnations of our universe) "come and go." Several other important results emerge from God Theory such a functionally triune God. The Unified SuperStandard Theory has many other important parts described in the Current Edition of *The Unified SuperStandard Theory* and expanded in subsequent volumes.

Blaha has had a major impact on a succession of elementary particle theories: his Ph.D. thesis (1970), and papers, showed that quantum field theory calculations to all orders in ladder approximations could not give scaling deep inelastic electron-nucleon scattering. He later showed the eigenvalue equation for the fine structure constant α in Johnson-Baker-Willey QED had a zero at $\alpha = 1$ not 1/137 by solving the Schwinger-Dyson equations to all orders in an approximation that agreed with exact results to 4th order in α thus ending interest in this theory. In 1979 at Prof. Ken Johnson's (MIT) suggestion he calculated the proton-neutron mass difference in the MIT bag model and found the result had the wrong sign reducing interest in the bag model. These results all appear in Physical Review papers. In the 2000's he repeatedly pointed out the shortcomings of SuperString theory and showed that The Standard Model's form could be derived from space-time geometry by an extension of Lorentz transformations to faster than light transformations. This deeper space-time basis greatly increases the possibility that it is part of THE fundamental theory. Recently, Blaha showed that the Weak interactions differed significantly from the Strong, electromagnetic and gravitation interactions in important respects while these interactions had similar features, and suggested that ElectroWeak theory, which is essentially a glued union of the Weak interactions and Electromagnetism, possibly modulo unknown Higgs particle features, be replaced by a unified theory of the other interactions combined with a stand-alone Weak interaction theory. Blaha also showed that, if Charmonium calculations are taken seriously, the Strong interaction coupling constant is only a factor of five larger than the electromagnetic coupling constant, and thus Strong interaction perturbation theory would make sense and yield physically meaningful results.

In graduate school (1965-71) he wrote substantial papers in elementary particles and group theory: The Inelastic E- P Structure Functions in a Gluon Model. Phys. Lett. B40:501-502,1972; Deep-Inelastic E-P Structure Functions In A Ladder Model With Spin 1/2 Nucleons, Phys.Rev. D3:510-523,1971; Continuum Contributions To The Pion Radius, Phys. Rev. 178:2167-2169,1969; Character Analysis of U(N) and SU(N), J. Math. Phys. <u>10</u>, 2156 (1969); and The Calculation of the Irreducible Characters of the Symmetric Group in Terms of the Compound Characters, (Published as Blaha's Lemma in D. E. Knuth's book: *The Art of Computer Programming Vols. 1 – 4*).

In the early 1980's Blaha was also a pioneer in the development of UNIX for financial, scientific and Internet applications: benchmarked UNIX versions showing that block size was critical for UNIX performance, developing financial modeling software, starting database benchmarking comparison studies, developing Internet-like UNIX networking (1982) and developing a hybrid shell programming technique (1982) that was a precursor to the PERL programming language. He was also the manager of the AT&T ten-year future products development database. His work helped lead to commercial UNIX on computers such as Sun Micros, IBM AIX minis, and Apple computers.

In the 1980's he pioneered the development of PC Desktop Publishing on laser printers and was nominated for three "Awards for Technical Excellence" in 1987 by PC Magazine for PC software products that he designed and developed.

Recently he has developed a theory of Megaverses – actual universes of which our universe is one – with quantum particle-like properties based on the Wheeler-DeWitt equation of Quantum Gravity. He has developed a theory of a baryonic force, which had been conjectured many years ago, and estimated the strength of the force based on discrepancies in measurements of the gravitational constant G. This force, operative in D-dimensional space, can be used to escape from our universe in "uniships" which are the equivalent of the faster-than-light starships proposed in the author's earlier books. Thus travel to other universes, as well as to other stars is possible.

Blaha also considered the complexified Wheeler-DeWitt equation and showed that its limitation to real-valued coordinates and metrics generated a Cosmological Constant in the Einstein equations.

The author has also recently written a series of books on the serious problems of the United States and their solution as well as a book on the decline of Mankind that will follow from current social and genetic trends in Mankind.

In the past twenty years Dr. Blaha has written over 80 books on a wide range of topics. Some recent major works are: *From Asynchronous Logic to The Standard Model to Superflight to the Stars, All the Universe!, SuperCivilizations: Civilizations as Superorganisms, America's Future: an Islamic Surge, ISIS, al Qaeda, World Epidemics, Ukraine, Russia-China Pact, US Leadership Crisis, The Rises and Falls of Man – Destiny – 3000 AD: New Support for a Superorganism MACRO-THEORY of CIVILIZATIONS From CURRENT WORLD TRENDS and NEW Peruvian, Pre-Mayan, Mayan, Anatolian, and Early Egyptian Data, with a Projection to 3000 AD,* and *Mankind in Decline: Genetic Disasters, Human-Animal Hybrids, Overpopulation, Pollution, Global Warming, Food and Water Shortages, Desertification, Poverty, Rising Violence, Genocide, Epidemics, Wars, Leadership Failure.*

He has taught approximately 4,000 students in undergraduate, graduate, and postgraduate corporate education courses primarily in major universities, and large companies and government agencies.

He developed a quantum theory, The Unified SuperStandard Theory (UST), which describes elementary particles in detail without the difficulties of conventional quantum field theory. He found that the internal symmetries of this theory could be exactly derived from an octonion theory called QUeST. He further found that another octonion theory (UTMOST) describes the Megaverse. It can hold QUeST universes such as our own universe. It has an internal symmetry structure which is a superset of the QUeST internal symmetries.

Recently he developed Octonion Cosmology. He replaced it with HyperCosmos theory, which has significantly better features. He developed a fractionalization process for dimensions, particles and symmetry groups. He also described transformation that reduced particles and dimensions to a far more compact form. He also developed a precursor theory ProtoCosmos that leads to the HyperCosmos.

The author showed that space-time and Internal Symmetries can be unified in any of the ten HyperCosmos spaces in their associated HyperUnification spaces. The combined set of HyperUnification spaces enable all HyperCosmos dimensions to be obtained by a General Relativistic transformation from one primordial dimension in the 42 space-time dimension unified HyperUnification space.

At present the author devel;oped the Cosmos Theory that incorporates ProtoCosmos Theory, HyperCosmos Theory, Limos Theory, Second Kind HyperCosmos Theory and HyperUnification Spaces. He has introduced PseudoFermion wave functions and theory, He has related Cosmos Theory to Regge trajectories of spaces, parton theory, Veneziano amplitudes, Fibonacci numbers and Ramsey numbers. He has calculated an approximation to the difficult R(n,n) Ramsey numbers.

He has developed a Gambol Model that successfully accounts for e-p deep inelastic scattering, fundamental particle resonances, hadron scattering, and the inner structure of particles based on confinement through Casimir forces of ideal gambol gases. The Gambol Planckian Distribution was derived.

He has applied the Gambol Model to particles, universes, and the Cosmos of universes. He showed that the Cosmos may have a distribution of 23 universes corresponding to various Cosmos spaces.

Recently he showed that Cosmos Theory follows from the number of independent asymmetric tensors in a dimension r. He also showed the close parallel between the form of γ-matrices and Cosmos Theory dimension arrays. The closeness suggested that dimension arrays have the same importance as γ-matrices for fermions.

He demonstrated that the pressure of fermions within a space of dimension r balances the Casimir vacuum energy force for 18 dimensions. He showed that $2e\pi = 17.02$ marks the critical point where pressure balances Casimir force, which implies $r = 18$ is the highest dimension Physical Cosmos space. The dimension $2e\pi$ appears to set the approximate dimension for Cosmos spaces with dimension array size $2^{r+4} \cong (17.02/8)^{r+4} . \cong (e\pi/4)^{r+4} . \cong 2.13^{r+4}$.

Now he has found the sequence of Coupling Constant values in the Standard Model and UST.